A TRAVELLER'S
LIFE

Eric Newby

A TRAVELLER'S
LIFE

LITTLE, BROWN AND COMPANY
BOSTON   TORONTO

Contents

Contents

Illustrations

To My Fellow Traveller

Introduction

This book is not an autobiography. It concerns itself for the most part, as the title suggests, with my life as a traveller in however modest a fashion from the time I was born more than sixty years ago.

Some of these travels were in distant places, in what used to be referred to as 'foreign parts'. But this is by no means true of all of them, and some of them were very near home indeed, for I agree with Ogden Nash's more or less unassailable definition of what constitutes a foreigner and what is a foreign part:

> The place you're at
> Is your habitat.
> Everywhere else you're a foreigner.

If you can bring yourself to believe this, it takes a lot of the sting out of the cost of travel; and it is why I felt it reasonable to include my journeys through Harrods – a strange early adventure which befell me and my somewhat oversexed nurse while she was propelling me in a baby carriage through a London suburb – as well as an account of some equally bizarre excursions into the underworld of the London sewers by night while working as a fashion buyer of dresses, retailing at ten guineas and upwards, for a chain of department stores during the day.

The somewhat episodic nature of the book is because one cannot continue going round the world for ever without intermissions in which one tries to make money, licks one's wounds, and re-equips oneself for further ventures. Even a traveller such as the Arab Ibn Battuta, born at Tangier in 1304 – perhaps the greatest traveller of all time who, in the course of his life, was estimated to have covered seventy-five thousand miles not counting detours, the only medieval traveller who is known to have visited the lands of every Muhammedan ruler of his time, quite apart from such infidel countries as Ceylon and China – was not always on the go, taking time off to get married here and there or to act as a counsellor of moderation to a mad potentate. In fact, travellers such as those who go into orbit and fail to come out of it, or travellers like the Jew who spat at Christ at the crucifixion and was condemned to wander the world for ever, can only be regarded as exceptionally unfortunate.

In his writings, the Venerable Bede compared the span of human life to coming out of darkness into a lighted hall and, having reached the end of it, finding oneself under the necessity of setting off once more into the all-embracing gloom. To me life has been more like one of those sections of *autostrada* on the Italian Riviera, on which there are lots of tunnels, some long, some short, with sunlit open spaces of varying lengths between them for which the darkness leaves one temporarily dazzled and often unprepared.

Why do people travel? To escape their creditors. To find a warmer or cooler clime. To sell Coca-Cola to the Chinese. To find out what is over the seas, over the hills and far away, round the corner, over the garden wall – with a ladder and some glasses you could see to Hackney Marshes if it wasn't for the houses in between, in the words of the old music hall song, the writer of which one feels was about to take off.

Why have I travelled? Difficult to answer, that is when not engaged in the equivalent of selling Coca-Cola to the Chinese (large size dresses in Leeds), or travelling as a sailor or a soldier. Partly, undoubtedly, for amusement and sheer curiosity and partly, as Evelyn Waugh wrote in the preface to a book I wrote which described a journey through the Hindu Kush, to satisfy 'the longing, romantic, reasonless, which lies deep in the hearts of most Englishmen, to shun the celebrated spectacles of the tourist and, without any concern with science or politics or commerce, simply to set their feet where few civilized feet have trod'.

I

Birth of a Traveller
(December 1919)

BIRTHS, MARRIAGES, DEATHS

BIRTHS

NEWBY. – On the 6th December at 3 Castelnau Mansions, Barnes, SW13, to Hilda Newby, wife of Geo. A. Newby – a son.

In this extravagant fashion – altogether it cost 50p ($1.95), at a time when Lady Secretaries with shorthand and typing were earning around £3.50 ($13,65) a week – my arrival was announced on the following Tuesday, 9 December, in *The Times* and the *Daily Telegraph*, two of the daily newspapers my father 'took in' at that period. The other was the *Daily Mirror*, then a rather genteel paper, which he ordered for my mother, but never looked at himself, and which she passed on to the cook/housekeeper when she had finished with it. From then on it was also passed on to a nurse.

As an event my birthday can scarcely be said to have been one of great consequence except to my parents, their relatives and friends. What is perhaps more interesting, and I hope the reader may think so too, is what sort of day that now far-off Saturday in December 1919 turned out to be, and what was going on in the world beyond the windows of that first-floor flat in which I was born facing the Metropolitan Waterboard's reservoirs and filter beds by the Thames on the Surrey side of Hammersmith Bridge.

At 3.45 a.m., the ghastly hour I chose, or rather the doctor chose, for my arrival – I had to be hauled out by the head – conditions must have been pretty beastly in Barnes. It was a dark and stormy night, with a fresh wind from the west whose gusts would have been strong enough to blow clouds of spray

* All sums of sterling have been converted from pounds, shillings and pence to their decimal equivalents and to American currency.

from the big reservoir (which was opposite our flat by the bridge and which has now been filled in to make playing fields for St Paul's School) over the pavement and right across the main road (which was called Castelnau but which all the inhabitants knew and know to this day as Castlenore) as it always did when the wind was strong from that particular quarter, sometimes, but rarely at 3.45 a.m. wetting unwary pedestrians and people travelling in open motor cars.

And it was certainly dark, although the moon had been up for more than thirteen hours and was only a day off full. It would be nice, more romantic, altogether more appropriate for a potential traveller, to think of myself arriving astride the Centaur, and, Sagittarius being in the ascendant, perhaps carrying the latter's arrows for him, as we moved across a firmament in which ragged clouds were racing across the path of a huge and brilliant moon; but it was not to be. It was ordained that I should be a child not only of darkness but of utter darkness, of ten-tenths cloud.

It was not much of a night for the distinctly grumpy, and, from what I subsequently gathered from my mother, very pompous Harley Street gynaecologist to be out, summoned from his residence in Hampstead to this distant and un-fashionable address at 2 a.m. by my father, using the telephone which he had had installed expressly for this contingency. On the other hand, it could have been colder. At 4 a.m. the thermometer at Kensington Palace, a couple of miles away on the other side of Hammersmith Bridge, registered a temperature well above freezing in the 40s Fahrenheit.

This 'specialist' subsequently performed an operation on my mother so incompetently and, so far as there was any possibility of her having any more children, so definitively, that years later the operation became the subject of a highly critical article in one of the medical periodicals in which my mother was referred to as Mrs N and the surgeon as Mr X, by which time he was dead and beyond the processes of the Law. That night he travelled to Barnes in what my father described to me when I was old enough to be curious about the circumstances of my birth as 'an electric brougham'.

The driver of such a machine sat outside, perched high up on a box, fully exposed to the elements – which would have been necessary if he had been driving a horse – while the passengers were accommodated in its leather upholstered and buttoned interior in considerable comfort. And, in fact, the effect produced by one of these contraptions, which looked as if its horse had bolted without it but was still moving forward by the force of gravity, steered by its gloomy, peak-capped driver with a wheel on a vertical column (gloomy because for most of his life he had probably driven horse-drawn broughams and regarded this development as an affront to nature), was highly comical. And when I eventually travelled in an electric brougham, aged five, on a night of thick pea soup fog and torrential rain in December 1924, from my grandparents' house in Winchester Street, Pimlico, back to Hammersmith Bridge, with my mother and father and an uncle and aunt, all warm and dry and full of food inside, while the driver was drenched with rain and half asphyxiated by fog on the outside, I laughed till I cried, all the way shrieking, 'He hasn't got a lid on!'

It was my mother, who was in a position to feel the full force of the specialist's grumpiness, who told me about it. However, why he was grumpy when he was being paid so highly for something he had contracted to do, *and* was treated to whisky on his arrival and champagne on my arrival, as well as chicken sandwiches, is not clear. Presumably the nature of his job must have accustomed him to working at odd hours, like lighthouse keepers and policemen. His grumpiness, however, was as nothing compared with that of another telephone subscriber to whose number my father was connected in error before getting through to the specialist, the work of one of the operators at the Hammersmith Exchange, whose fruity 'SORRRYY YOU'VE BEEN TRRROUBBLED!' did nothing to convince the wretched man, dragged from his bed at two in the morning, that anyone was sorry at all.

It would not have been much of a night for the homeless poor, their clothes stuffed with newspaper, who slept rough on the towing path down by the river all through my early childhood, and who would certainly have been there that night. Most of them were regulars. Some were terrifying-looking women; some were 'tramps', the first 'real' travellers I can remember seeing, pointed out by my nurse. But not many of them would have been tramps because most tramps were too solicitous of their personal comfort to share the appallingly draughty, unspeakably filthy but more or less rain-proof camping place used by these unfortunate outcasts, up against the reeking abutments under Hammersmith Bridge, only about fifty yards from where, a boisterous baby, I was now giving tongue. But on this particular night with a high spring tide some time after midnight (high water at London Bridge was at 12.12 a.m.) their pitch would have been a couple of feet under water for an hour or more, and they would have been sleeping among the bushes down towards Putney, or up against the trunks of the huge black poplar trees that grew along the towing path opposite Chiswick Mall further upstream to which normal spring tides did not reach.

Some of these men and women drank methylated spirits. If they became violent, they were 'taken into custody' by the police. This usually meant that a couple of unfortunate constables, sometimes one alone, had to strap the prisoner, male or female, who by this time would probably be striking out, biting and scratching, to a handcart and then wheel it a mile or more up Lonsdale Road from the Boileau Arms, which everyone called, and still calls, 'Ther Boiler', to Barnes Police Station with the occupant roaring loudly enough to wake the dead. If more than one person had to be taken into custody, a Black Maria was sent for.

When it dawned, the day was even more rumbustious than the night. And when the sun rose, just before eight o'clock, like the moon, it remained invisible. Thunderstorms visited many parts of the country, accompanied by hail, sleet or snow and west or north-westerly winds which reached gale force in high places. In Lincolnshire, the Belvoir Hunt, having 'chopped a fox' in Foston Spinney (seized it before it fairly got away from cover), 'were hunting another from Allington when scent was totally swept away by a tremendous rainstorm'.

'Flying Prospects' on my birthday were not good, according to *The Times*. It is now difficult to imagine that a pilot, or even a passenger, might actually buy a newspaper in order to find out whether it was safe to 'go up', but it must have been so, otherwise there would have been no point in publishing the information at all. 'Unsuitable for aviation or fit only for short distance flying by the heaviest sort of machine' was what the communiqué said. 'Sea Passages' were equally disagreeable. The English Channel was rough, with winds reaching forty miles an hour, and there was extensive flooding in France.

But if the weather was disturbed that Saturday, it was as nothing compared with the state of great chunks of Europe and northern Asia. In spite of the fact that the advertising department of *The Times* had chosen this particular Saturday to announce 'PRESENTS SUGGESTIONS FOR THE GREAT PEACE CHRISTMAS', on it Latvians were fighting Germans, on whom they had declared war a week previously on 28 November, and so were the Lithuanians. In Russia, on the Don and between Voronezh and Kirsk and in Asia, beyond the Urals, along the line of the Trans-Siberian Railway, where typhus was raging, Bolsheviks and White Russians were engaged in a civil war of the utmost ferocity. Meanwhile, that same Saturday, while their fellow countrymen were destroying one another, with their country in ruins and becoming every day more ruinous, Lenin and Trotsky and the 1109 delegates of the Seventh All-Russian Congress of Soviets passed a resolution to the effect that 'The Soviet Union Desires to Live in Peace with All Peoples'. On that day, too, Lenin told the Congress that 'Communistic Principles were being utterly disregarded by the Russian peasantry.'

That day, too, much nearer home, while I was taking my first nourishment, as it were, in the open air, French Army units with heavy guns were rumbling across the Rhine bridges in order to force the Germans to ratify the peace treaty which they had signed at Versailles in June; and in the same issue of *The Times* which carried the headline about 'THE GREAT PEACE CHRISTMAS', there were other headlines such as 'GUNS ACROSS THE RHINE' and 'WAR IMMINENT', although who was to fight another war with millions killed and wounded, armies in a state of semi-demobilization, and millions more dying or soon to die from sickness and starvation was not clear. Nevertheless, that weekend, the only thing, theoretically, that stood between the protagonists and another outbreak of war, was the Armistice, signed in a French railway carriage parked in a wood, thirteen months previously, so that, equally theoretically, it would simply have meant carrying on with the old one. That weekend, too, the Americans quitted the peace conference.

There was, altogether, a lot about death in the papers that Saturday. It was as if Death the Reaper, an entity embodied by cartoonists in their drawings as a hideous, skeletal figure, and it would have been difficult to have lived through the last five years without thinking of death as such, had become dissatisfied with his efforts, had once again sharpened his scythe and was already cutting fresh, preliminary swathes through the debilitated populations of the vanquished powers, as if the great influenza epidemic, which reached its peak in Britain in March 1919, and which altogether killed more people in Europe than all the shot and shell of four and a half years of war, had not been enough.

In Britain, that Saturday, things were rather different. Bank rate was six per cent, exports were booming. On Friday, the US dollar closed at $3.90 to the pound. The only disquieting news that morning, and that was more or less a rumour, was that there was a possibility of a number of pits being forced to close in the South Wales anthracite fields.

Altogether, for many people that Saturday, life seems to have gone on much as it had done before the Deluge. Giddy and Giddy, House Agents, offered a luxuriously furnished town house, facing Hyde Park, with thirteen bed and dressing-rooms for £26.25 ($102.40) a week. Harrods announced Laroche champagne, 1911, the last vintage generally available (shipped) since the war, at £6.50 ($25.35) a dozen. Very old vintage port (Tuke Holdsworth) was £4.50 ($17.55) a dozen. Not advertised in *The Times* or the *Daily Telegraph*, but still listed in Harrods' enormous current catalogue, (and for some years to come) under 'Livery', were red plush breeches for footmen.

Domestic servants were still comparatively inexpensive, although more difficult to find, than they had been before the war. That Saturday Lady Baldwin, of 37 Cavendish Square, advertised for a housemaid, 'five maids and a boy kept, wages £28-£30 ($109-$117) a year'. And there were vacancies for live-in under nurses, at £25 ($97.50) a year, the price of a high-class baby carriage of the sort that my mother had acquired for me.

That Saturday, too, wholesale garment manufacturers, at what was, and still is, known as 'the better end of the trade', the sort of firm my father was a partner in, were advertising jobs in their workrooms for bodice and skirt makers at around £2.50 ($9.75) for a five and a half day, forty-nine hour week (8.30 a.m. to 5.30 p.m. week-days, 8.30 a.m. to 12.30 p.m. Saturdays), £130 ($507) a year, which made the 50p spent on announcing my birthday seem hideously extravagant.

That Saturday some London fashion houses, including the then ultra-fashionable Lucile, in Hanover Square, were advertising for 'Model Girls', in emulation of Paul Poiret, the Parisian designer, who had just returned from the army and for the first time showed clothes in living models.

A sketch in *The Times* that Saturday shows that clothes were good-looking, if not positively saucy. Dresses, according to their fashion correspondent, were '*décolleté*, sometimes dangerously low', in brilliant colours, with tight, mid-calf-length skirts. Jet was high fashion for the evening: embroidered on coloured velvet, used for making girdles and shoulder straps. Feathers, which had been used for years for making headdresses for evening, were being replaced by flowers, 'as little like nature as possible?', although another couple of years were to pass before the Importation of Plumage (Prohibition) Act became law. The ultra-fashionable were already wearing the long, skimpy jerseys which were to become a sort of hallmark of the 1920s; but there was nothing about them in the papers the day I was born.

Yet in spite of all this display of what an American politician described as 'normalcy', 'The Great War', as it would still be referred to by the British far into the next one, although over, must have seemed terribly close to most people, as it still must do today to anyone reading some of the classified advertisements which appeared in the quality papers that Saturday. The request for a lady or gentleman to play once a week at a *thé dansant* in a hospital for shell-shocked officers. The offers to keep soldiers' graves trimmed and lay headstones in the neighbourhood of Albert, Bapaume and Péronne – the dead had not yet been gathered together in communal cemeteries. The endless columns of advertisements inserted by ex-servicemen, under

'Situations Wanted' (there were 350,000 of them unemployed), part of the huge citizen army of the still living that was being demobilized into a world in which, in spite of there being whole generations of dead, there was not enough work for all. Such advertisements, inserted by ex-officers, warrant officers, petty officers, NCOs and men of superior education (the labouring classes did not advertise their services in this way), were some of them despairing, some of them pathetic, some of them hopeless:

Ex-Service Man. Loss of right arm, seeks situation as Window Dresser or Shopwalker.

Demobilized Officer. Aged 21, *4½ years' service* [my italics]. Good education. Left school to join up, therefore no experience. Accept small salary until proficient.

Will anyone lend Demobilized Officer, DSO, just starting work again, £5000 ($19,500) for one year? Highest references. Applicant desperately pressed by money-lenders. No Agents. Write Box J.28.

Money-lenders were so numerous that they had whole classified sections to themselves. Most of them offered 'immediate advances on note of hand alone'. Their advertisements make repulsive reading, even across such a gulf of years.

A far more prominent advertisement than any of these announced the setting up of what was called the Bemersyde Fund, opened by the Lord Mayor of London and the Right Honourable Lord Glenconner, 'to acquire the estates of Bemersyde from its owner and have the same conveyed to Field-Marshal Earl Haig, a member of the well-known whisky distilling family, as a personal gift from the people of the British Empire – the consideration for the purchase being £53,700 ($209,430)'.

Altogether – leaving present for the headmaster (the Estates of Bemersyde), although he had not been a very good headmaster, the boys (or what was left of them, for it had been rather a rough school with a lot of mud in the playing fields) now going out into the world to seek their fortunes – there was a distinctly end-of-term feeling in the air. But in spite of this there was no singing of 'Lord Dismiss Us With Thy Blessing' as one would perhaps expect on such occasions and as there was at the schools I later attended. Possibly because the only songs the boys knew were not hymns but songs that had become dirges: 'Pack Up Your Troubles', 'Tipperary', 'It's a Long Long Trail A'Winding' and 'I Don't Want To Join The Army'.

Even the Ministry of Munitions and the Admiralty were selling up. That day and every day there were offers for sale by auction of aerodromes, enormous munition factories, equally enormous hutted camps, and of minesweepers, motor chara-bancs, fleets of ambulances, ships' boilers, railway engines, enough barbed wire to encircle the earth, millions of cigarettes in lots, miles and miles of ships' hawser, inexhaustible supplies of bell tents, cereal ovens, lower fruit standard jam, wicker-covered stoneware jars, torpedo boats, with and without engines and part-worn and unworn issue clothing, etc., etc., etc., so inexhaustible that many items were still being sold off twenty years later on the eve of the next world war, when the whole stocking up process began all over again. Everything, except projectiles and the means of discharging them, was open to offer and even these would eventually come on the market, but for export only.

In fact the world was changing with a rapidity that would have been unbelievable in 1914, even though it was still possible to buy red plush breeches for footmen and under nurses could still be acquired with comparative ease. Yet it was, one sees in retrospect, only a temporary acceleration. If it had continued at the rate envisaged in 1919 man would probably have stood on the moon by 1939.

That Saturday, if the weather had allowed, one could have flown to Paris or Brussels in one of the new Handley Page Commercial Aeroplanes, at a cost of £15 ($58.50) single fare, a service of which my parents availed themselves the following year when my father went to Paris to buy 'models' to copy from, amongst others, Poiret and Madame Vionnet, which were made to my mother's dimensions so that she could show them, or copies of them, in London. The fifteen or twenty passengers travelled at a speed of ninety miles an hour in a large saloon furnished with carpets, curtains, armchairs, clocks, mirrors, telephones and flower vases.

There was also news of the Aerial Postmen (in *The Times*),

who for the last fifteen weeks had been carrying mail between Hounslow and Paris, taking about two and a half hours. And there were confident predictions that morning of regular mail services to Madrid, Vienna and Rome, and even further afield: to Cairo in twenty-three hours, a journey which then took four days; New York in forty-seven and a half hours instead of five and a half days (Alcock and Brown had succeeded in flying the Atlantic nonstop from Newfoundland to Ireland in June); Tokyo in sixty-six hours, instead of fifteen days; and even London-Sydney in an estimated hundred and twelve hours, against a month.

This was no madman's dream. Even while the readers were digesting this information that Saturday, Captain Ross Smith landed safely in West Java, while on what was to be the first flight from London to Port Darwin, which he reached on 10 December, having covered 11,294 miles in 668 hours 20 minutes, just under twenty-eight days.

Even more incredible, especially to older readers, must have been the realization that the widespread use of the internal combustion engine was not just a phenomenon of war, and that for all practical purposes horse-drawn vehicles were doomed. If they did not believe the evidence of their own eyes, when now long ago they had seen, for example, the first horse-drawn brougham converted to run on electricity, it was only necessary for them to glance through some of the classified advertisements in the newspapers under 'Horse and Carriages':

'ALDRIDGE'S, ST MARTIN'S LANE. LONDON. ESTABLISHED 1753. On Wednesday, 10 December. Well known stud of horses, newspaper vans, harness and stable sundries, the property of the *Star* newspaper, who are discontinuing their horse department and adopting motor transport entirely.'

'ELEPHANT AND CASTLE HORSE REPOSITORY ... Motor Auction Sales every Thursday at eleven o'clock.'

That Saturday while I lay in my nursery in SW13, tucked up in a bassinet, which was shrouded superfluously in voile to keep off any stray draughts that might conceivably be about, and with a good coal fire burning in the grate, a number of totally unconnected events occurred and were reported in *The Times* the following week.

That day the Bishop of Oxford confirmed more than two hundred boys at Eton; ten people were injured in a tram accident in Hackney; Field-Marshal Sir Evelyn Wood was buried with full military honours at Aldershot; four thousand members of the United Garment Workers Union met in the Mile End Road to demand a forty-eight hour week; and at a congregation held at Cambridge, the Vice Chancellor presiding, a proposal that a syndicate be appointed to consider whether women students should be admitted to membership of the University and, if so, with what limitations, if any, was carried without opposition.

That day also, Renoir, who had died three days previously at the age of seventy-eight, was buried; the entire staff of the Army and Navy Stores which had been on strike went back to work*; ex-soldiers at Bangor Training Centre completed a pair of shooting-boots for the Prince of Wales; the body of a young woman, wearing a velvet blouse, dark skirt and patent leather boots, was washed up by the tide at Swansea; Florence Langridge was sentenced to three months for giving Harrods a dud cheque, while masquerading as the widow of a captain of Hussars; William Docker, nineteen years a railway shunter with the Great Western Railway Company, more harshly dealt with, was given three months with hard labour for stealing a dozen pairs of stockings from the company; Stoke Poges beat Oxford University at golf; the will of Henry Clay Frick was published – he left $145,000,000 (£3,411,764) bequeathing all but $25,000,000 (£470,588) to educational and philanthropic objects.

That evening the members of the Overseas Club entertained the staff at a fancy dress dance, a function at which I would, in

---

* Up to this time employees at the Army and Navy Stores had worked a sixty-hour week, and men of sixty were being paid as little as £2-£2.25 ($7.80-$8.80) a week . In one department fifty women were employed whose average wage was less than £1.26 ($4.90) a week. The strikers who received much sympathetic support from the public and the newspapers were successful in improving their lot.

retrospect, have dearly liked to have been present. Just as I would also have liked to have been present at Diaghilev's productions of *La Boutique Fantasque*, *The Three Cornered Hat* and *Midnight Sun* at the Empire Theatre, with Karsavina, Tchernicheva, Massine and Svoboda; although as a future traveller I might have been expected to derive more benefit from attending Lowell Thomas's *With Allenby in Palestine* and *With Lawrence in Arabia*, – 'Two entertainments for the price of one' – at the Albert Hall; or a showing of *Tarzan of the Apes* at the New Gallery; or attending a lecture on The Antarctic Expedition of 1914–17 by Sir Ernest Shackleton at the Hampstead Conservatoire.

That day, too, the Royal Mail Steam Packet Ship, *Bogota*, left the Thames on the top of the tide, bound for Valparaiso.

That afternoon the muffin man ringing his bell went down Riverview Gardens, the side road outside 'Ther Mansions' as the local tradesmen who dealt with my mother called them, in which there were other blocks of flats, carrying his muffins in a wooden tray covered with a green baize cloth, which he balanced on his head; and the lamplighter came and went on his bicycle (lighting-up time that evening was 4.21 p.m.), lancing the gas lamps in the street into flame with a long bamboo pole.

I did not know about any of these exciting things, and if I had I would not have cared. I had no intention of going anywhere, certainly not to Valparaiso in the SS *Bogota*. And so ended my birthday. For all concerned it had been a jolly long one.

2

The Baby as a Traveller

At the time I was born, and for long afterwards, 'middle-middle-class babies', of whom I was one, rarely travelled in motor cars, 'middle-middle-class motors' being mostly open ones, and sometimes difficult to close if a change of weather demanded it. When I went on holiday the year after I was born, and the year after that – and photographs assure me that I did – it was by train from Victoria, Waterloo, or Liverpool Street with lots of trunks and my vast £25 ($97.50) pram with its fringed awnings and a sort of shotgun holster for parasols or umbrellas, according to what was going on overhead, which needed a couple of porters to lift it into the guard's van, which meant lots of lovely tipping. In those years I went to nice, unadventurous places such as Frinton, Bembridge, Broadstairs, or Cliftonville which was ideal for babies because there nature had been almost completely eradicated. I cannot remember the lot, for I spent a week here, a week there, presumably as the spirit moved me.

Down on the beach at one or other of these or similar resorts, surrounded by babies of similar age and condition (*The Times* recorded some nineteen babies as having been born on the same day I was), I used to pass the cooler days at some of them against a background of cliffs and the only recently outmoded, horse-drawn bathing machines which, horseless, rather like the electric brougham but without the electricity, still performed the function for which they had been built but were now parked permanently above high-water mark. If the temperature rose above 55° Fahrenheit, an admittedly rare occurrence at the seaside (which is also the correct temperature for serving draught beer in Britain), we would all insist on being taken indoors and placed in our bassinets, still swathed in voile, not as a protection against treacherous currents of air but, on the same principle as Bedouins swathe themselves in wraps, against the intense heat.

The truth is that babies do not like travel, and I was no exception. Babies are unadventurous. Babies act as grapnels to prevent 'the family' dragging its ground. That is why they were invented. Perversely, their desire for fresh horizons comes much later when they have already begun to 'attract' fares, and can no longer travel free; by which time they are no longer babies at all.

The prospect of the Great Glen, the Grand Canyon, the wastes of the Sahara at sunset, the entrancing, set-piece landscapes of Tuscany, all leave them equally indifferent, and usually breaking out in a rash which takes weeks to clear up. Useless to consult the baby about where it would be prepared to go without these alarming side effects because it will never express an opinion until it arrives at its destination, when it is invariably adverse.

Thus did I spend my first two years of travel. It is a wonder, and a credit to my parents' resilience, that I did not succeed in driving them permanently round the bend.

What can one truly remember of one's infant life when one comes to write about it years later, putting as it were one's hand on one's heart, separating in the mind's eye what one can really remember from what one has been told, separating fact from fiction, or what is more factual from what is more fictional on those frontiers where these nuances become blurred and indistinct?

In the case of my own childhood this *mélange* of what I could really remember and what I thought I could remember was the result of looking over a long period of years at hundreds of photographs made with a 3A Eastman Kodak. Some of them were taken in what even today would be regarded as technically difficult circumstances, such as foreground figures photographed against shimmering summer seas, long before exposure meters came into use. In many of them I was either the principal figure or, if it was a landscape, was somehow or other included as an extra.

Thus I appear, embalmed as it were, in volume after volume of now fragile cloth and morocco-bound albums, most of them with the relevant dates and places written neatly above them in ink: in the pram at Frinton, Whitsun 1920; on the sands below the white cliffs at Broadstairs, facing the English Channel, in front of a striped bathing tent with my father's white buckskin shoes parked outside it – he may have gone for a dip – ensconced on a cushion on a deck-chair like an infant Dalai Lama, August 1920; barely able to stand, supported by my mother like a drunken man, wearing a white woolly suit and defiantly waving a rattle, behind the privet hedge in the front garden of Three, Ther Mansions, on a bleak day in March 1921; apparently alone at Bembridge, Isle of Wight, apart from a girl in a gym smock who is 'bothering me', September 1921; wearing a floppy white sun hat and rubber waders, digging away on the beach at Bournemouth with a wooden spade and, without the waders, riding on a donkey outside a subscription library on the front, Whitsun 1922; on the Isle of Wight again, this time in the side-car of a motor-cycle combination with my mother at the helm; on the rocks and in the bracken on Sark, July 1923.

How few other holiday-makers there were on the beaches, even in high summer in these years immediately after the war, is shown in those early photographs. At that time only the well-off went to the sea for a fortnight or a month. The great majority, that is of those who went away at all, went on day excursions as 'trippers'.

According to these photographs everywhere we went we must have picnicked. In every picture of a picnic a large wicker basket that would have needed two people to carry it, loaded with mounds of food, and batteries of Thermos flasks in their own special wicker containers, stand between us and whoever is taking the photograph.

One of these picnic photographs, taken in September 1921, shows my mother and I in a lane in Surrey, not far from the London to Portsmouth Road. It is a sunless, autumnal day, mist is beginning to rise from the fields beyond the hedgerow gate where our picnic has been set out, and by the roadside stands our splendid, shiny, open Napier motor car, the sort of motor car which Mr Toad would have planned to make off with if he had ever set eyes on it.

Although I remember the Isle of Wight as the place where I

first sat in the side-car of a motor cycle, at Easter 1923, much more I remember it as being the Place Where God Lived, although this was later, some time in the summer or autumn of 1925. It must have been during one of those interpolated holidays my mother was so adept at arranging at an instant's notice if my father had to go abroad without her, on the grounds that a change of air would do me good. He often used to go to Holland to sell enormous coats and costumes to the Dutch. With her she took her sister, my Auntie May, who loved travel, however banal.

On one occasion we made an excursion to a place near the middle of the island and some time in the afternoon of what I remember as a very hot day we arrived at our destination, a village of thatched houses that were clustered about the foot of a green hill, on the summit of which stood what seemed a very small church.* From where we stood it was silhouetted against the now declining sun, the rays of which shone through its windows, producing an unearthly effect.

There was no time to climb the hill to the church and have tea as well. If there had been, I am sure that my mother and my aunt, both of whom were interested in 'old things', would have done so. Instead, we had the tea, in the garden of one of the cottages, and while we were having it I heard my mother and my aunt talking about the place and how nice it was, which they called Godshill.

I was very excited. Godshill. If this was Godshill then God must live on it. God to me at this time and for long years to come

---

* It was originally intended that the church should be built at the foot of the hill near the site of the present village. However, when work was begun on it, the plan was vetoed by a band of local fairies. As a practical expression of their objection whenever the walls reached a particular height they proceeded to knock them down and carry the stones up to the top of the hill where they rebuilt the walls, after which they danced round them in a ring. After this had happened three times, the workmen who had on each occasion been forced to demolish the walls, carry the stones back down the hill and then build them up again in the low ground, lost heart and decided to build the church where the fairies wanted it to be built. As a result of this wise decision there was much jubilation among the fairies and when the church was finally completed they held a great *fête* on top of the hill to celebrate their victory, the sounds of their revelry being audible at a considerable distance.

was a very old, but very fit, version of Jesus and much less meek-looking. He had a long white beard, was dressed in a white sheet and was all shiny, as if he was on fire. He also had a seat in the front row of the dress circle, as it were, so that he could see immediately if one was doing wrong. This was the God to whom I prayed each night, either with my mother's help or with whoever was looking after me.

'Does he live on it?' I asked my mother.

'Yes,' said my mother, 'that's where he lives, darling, on top of the hill.'

I was filled with an immense feeling of happiness that this radiant being, whom I had never actually seen but who was always either just around the corner or else hovering directly overhead but always invisible, should live in such a shining, beautiful place; and I asked if we could climb the hill and see him. Unfortunately, the train was due and we had to hurry to the station. I cried all the way to it and most of the way back to Bembridge. I never went back to Godshill and I never will.

I can remember, in July 1923, being carried high on my father's head through the bracken in the combes that led down to the beaches on Sark, and once having reached them I can remember falling down constantly on the rocks and hurting myself, I considered, badly. And it was on Sark that I had my first remembered nightmare, in the annexe to Stock's Hotel, a charming, ivy-clad, farmlike building. I awoke screaming in what was still broad daylight with the sun shining outside my first-floor room in which the blinds were drawn, to think myself abandoned to a dreadful fate by my parents who were dining only a few feet away in the hotel, certain that I had 'gone off' to sleep. It was a nightmare of peculiar horror, because it was founded on fact; so horrible and at the same time so difficult to explain to anyone that for years I dared not confide the details to anyone, and to my parents I never did, although it recurred throughout my childhood, together with an almost equally awful one about falling down an endless shaft.

3

Rings Around the Tombs in SW13
(1923)

This hideous dream I last dreamt, after an interval of fifteen years, while escaping from the Germans in Italy in the autumn of 1943. It derived from an incident that occurred in the spring or early summer of 1923, the same year that we went to Sark. This incident took place in Barnes while I was on an outing with my nurse in what used to be called a mail cart or Victoria carriage. A mail cart was a machine made for the conveyance of children who have outgrown their prams, as I had, but were still unable to cover long distances on foot, bearing the same relation to a push chair as a Hispano-Suiza to an Austin Seven. In it the infant occupant sat upright with his back as it were to the engine, in this case whoever was pushing the thing. With the hood up conversation between pusher and pushed was precluded, unless the pusher stopped pushing and walked round to the front of the vehicle. It was in some ways a beautiful vehicle, the product of the pre-industrial revolution coach-builder's imagination and just as an electric brougham looked like a brougham that had lost its horse, so a mail cart looked like a Regency curricle which had lost its horses and was being pushed back to the stables by human hands.

My pusher was called Lily. She was my first and last real nurse. I can remember everything about Lily without the aid of photographs; but the photographs confirm that she was what I thought she was, even at that early age, a very good-looking in a soppy kind of way, raven-haired, distinctly friendly girl with dark rings round her black eyes. I have already referred to her in another book, *Love and War in the Apennines*, but she has to be resurrected yet again for the purpose of this narrative.

Lily had been kitted out by my mother in what must have been a moment of social aspiration in full nurse's rig. The winter outfit, navy-blue coat and a sort of pork-pie hat to match which she wore at a jaunty angle, was innocuous enough but the summer one was very different. It consisted of a short-sleeved blue-denim dress with starched white collar and cuffs, black silk stockings, high heels and a headdress made up of swathes of dark-blue veiling. Dressed in this outfit, a model girl's idea of a W1 or SW1 nurse, with the veiling and the black-rimmed eyes, she looked like a mixture of a houri and nurse in a blue film. In London, W1, or SW1, where nurses, in fact, tended to be rather plain, if not hideous or of forbidding demeanour, she would have been very conspicuous and they would probably have driven her from Hyde Park, if she had attempted to enter it, into that desert where nurses whose charges did not appear in Debrett were sent to languish, Kensington Gardens. In Barnes, SW13, the total effect of the uniform, Lily and her soppy, friendly air could have been nothing less than inflammatory. I loved Lily but even then at that tender age I recognized that it was in a different way from anyone else who ever looked after me; and I think Lily loved me, but in a different way from the way in which I loved her. Thus, because of all this, in her company, as a sort of accomplice or accessory after the fact, because I could easily have told my mother what was going on, I found myself being trundled to assignations, only one or two of which I can remember fully, with what I recall as old men (which meant that they might have been twenty years old) and my mother recalled years later when I was fully grown as 'dirty old men' (which probably meant that they were over forty).

The venues for these presumed encounters, for I never remember seeing any actual goings-on, were the towing path above Hammersmith Bridge near Chiswick Ferry which was grassy and on which a number of bushes grew, and a creepy and now desecrated and presumably deconsecrated cemetery on Barnes Common. In it Lily kept me quiet while, again presumably she made rings around the tombs, by giving me handfuls of Carrara marble and other more brightly coloured chippings to play with. Some of these tomb chippings found their way into my bath where they were discovered by my mother. Subsequent sleuthing led to Lily being surprised by my mother, whether while being about to 'do it', or while actually in the act of 'doing it', or simply being chatted up, whether on the towing path or in the cemetery or at some other trysting place, she never made clear. Whatever or wherever it was, Lily was instantly dismissed, although this was not until some time towards the end of 1924, the year following the events which I am now narrating.

Whether it was in pursuit of whatever she was in pursuit of, or we were simply on a new, adventurous walk, on the afternoon on which the happenings which led up to my nightmare took place, Lily pushed me in the mail cart up the towing path from Hammersmith Bridge as far as Chiswick Ferry. The ferry was for foot passengers only, and when it functioned at all, which was rarely, they were conveyed across the river by a ferryman in a rowing-boat. Having reached the ferry, as she usually did, Lily turned left down a narrow, unmetalled lane between two reservoirs from which it was separated by iron railings. This lane led to Lonsdale Road, the road up which the police used to push the drunk and disorderly on their handcart to Barnes Police Station. At Lonsdale Road she normally turned left for Hammersmith Bridge and home along the pavements. But on this particular day instead of doing this she crossed Lonsdale Road and continued to follow the alignment of the lane into what was, for me, unknown territory.

It was an eerie place. To the left of the lane, which was also unmetalled, a rather dreary expanse of fields with a farmhouse on the edge of it, what must have been one of the nearest farms to central London, stretched away towards the semi-detached developments that but for the war would have already engulfed them, as they would shortly. In these flat fields, some distance off, a line of what looked like men but I later discovered when I was older were rough-looking women wearing cloth caps and sacks in lieu of aprons, worked away, bent double among the vegetables.

To the right of the road a rusty corrugated-iron fence, its top cut into cruel, jagged spikes and festooned with brambles and old man's beard (an appropriate weed for Lily, perhaps, in the circumstances), separated it from the adjoining property, and along it a line of trees, possibly willows, with thick pollarded trunks grew, or rather rotted, for most of them were in the last stages of decay. The surface of the road was full of potholes with water in them, and in the ditches on either side was some of the detritus of civilization, what the French more expressively call *ordures* – broken lavatory pans, rusty oil drums, bits of bicycles and prams, broken shoes, awful items of discarded clothing, bundles of sodden newspaper, broken glass. It was certainly no place for a nanny and a small child in a mail cart. Some five years later, when I was at Colet Court (a London preparatory school), my favourite museum was the Imperial War Museum in South Kensington and there in the picture gallery I saw dozens of similar roads, only the potholes in the pictures were shell-holes and the trees had been shattered by gunfire, all painted by war artists on Flanders and other fields. It was therefore not surprising that when the fields were finally built over some years later and the lane became a respectable suburban road, whoever was in charge of naming roads in Barnes gave it the name it bears today, Verdun Road.

Against the largest and most decayed of these ruined trees a fire was burning, eating its way into the heart of it, and sitting close to the fire, although it was late afternoon it was still warm, were three of the hideous hags who, when the tide was right, slept up against the abutments under Hammersmith Bridge. And on the fire was an iron pot. It would have been impossible for anyone to say how old these creatures were. They were so blackened by smoke and smeared with filth that it was difficult to identify them as human beings. One of them was singing in a wild, tuneless mindless way and another was screeching at the third member of this ghastly triumvirate, while picking away like a monkey in her long, lank hair. The third one was tending the pot.

As we came abreast of them, the one who was looking for lice or nits in her companion's hair (for that is what she must have been doing), got to her feet and came towards us with surprising swiftness, with her horrible discoloured stockings dragging around her ankles, mumbling something about 'the baby' between her broken teeth. It was too much for me and I began to bellow; and it was too much for Lily who kicked up her heels and fled, pushing the mail cart through the water-filled potholes which she had previously carefully skirted, so that it bounced up and down on its springs, soaking herself in the process.

She did not stop until she reached the corner of Madrid Road where we were once again on a real, made-up road and enclosed by comforting suburbia. By this time she had more or less succeeded in calming me down.

'Horrible old thing,' she said, 'I thought she wanted to eat you up.'

And this not only set me off again but crystallized the dream so that it would always unfold in the same way: myself alone, forced by some irresistible power to walk along the lane with the sun sinking behind the corrugated-iron fence and the dying trees to the one where three cackling hags sit round a fire burning in the heart of it, preparing to make a cannibal feast of the infant Newby.

It was about this time that the tragic demise took place of Mrs George. Mrs George had been our cook/housekeeper since before I was born and it was to her that my mother used to pass on her copy of the *Daily Mirror* when she had done with it. When I was born she ceased to 'live in', arriving each morning before eight o'clock from where she lived, over the river in Hammersmith.

When she retired, early in 1923, she went to live in a house, so far as I can make out, in Glentham Road and continued to visit us. Glentham Road led down by what must have been one of the few hills in Barnes from Castelnau by the side of the reservoir from which the spray used to blow across the road. Mrs George was white-haired, fresh complexioned, large enough to qualify for one of the smaller sort of coat that my father sold to the Dutch, and motherly. Seen from the front, protected by an expanse of spotless, white starched apron she looked like a spinnaker that was drawing nicely. I loved Mrs George. She smelt lovely, of the things she was always baking and she let me help her to stir the Christmas pudding mixture which was delicious in its raw state but emerged from the oven in the form of puddings as heavy and black as cannon balls.

Mrs George called my mother 'Ther Missus' and my father 'Ther Master'. She called the enormous ochreous, to me rather creepy building at the bottom of Riverview Gardens with the words HARRODS FURNITURE DEPOSITORY written large on the side of it, 'Ther Suppository'.

Each week on her afternoon off Mrs George used to set off with her friend, another cook from round the corner, for Pontings store in Kensington High Street, always a magnet for domestics on their afternoons off, travelling on the No. 9 or 73 bus. With her, rain or shine, summer and winter, she always carried an umbrella and often, even when it was not raining, she used to be seen in the street with it up. This was her only eccentricity and no one will ever know why Mrs George took it into her head one day when the tide at Hammersmith Bridge was sufficiently low for her to go down some steps to the muddy foreshore and, fully clothed and with her umbrella up, although it was not raining, enter the water and be swept away by the still ebbing tide. It was not for lack of money. She was of a prudent nature. The coroner recorded a verdict of 'suicide while of unsound mind' which was more or less mandatory at that time.

'George gone,' I said when the news was eventually broken to me.

4

Travels in Harrods

As I indicated in an earlier chapter, my mother was a customer of Harrods before I was born. She had worked as a model girl in one of its fashion departments as long ago as 1912 and could probably have found her way around the place blindfolded. At the time she worked there it is unlikely that she was a model girl in the present sense of the word. Poiret, it is claimed, 'invented' them in 1919. Her job, or part of it, would probably have been to try on new stock when it came into the store so that the buyer, who at that time would have also been the department manager, or one of her deputies if they were inexpensive versions of 'models', could detect any defects which could give her the excuse, always a temptation if the buyer had over-bought, to send the garments back to the suppliers with a debit note. In the jargon this operation was known as 'passing'.

For those who have not read *Something Wholesale*, an account of my life with my parents in the garment industry, this would seem to be an appropriate moment to interpolate a little more information about my father.

My father was apprenticed to the drapery trade in 1887 at the age of thirteen, in the Brompton Road, where he slept under the counter of the shop, which was then commonplace. Later he graduated to the drapery department of Debenham and Freebody, which he left to become a partner in the firm of Lane and Newby, Mantle and Gown Manufacturers and Wholesale Costumiers, which was how the firm's letterheading described the scope of its activities well into the 1950s. He was an all-round sportsman, a pupil of Sandow, the strongest man in the world, who eventually destroyed himself by lifting an enormous motor car out of a ditch unaided. My father used to go down to Whitechapel to be 'pummelled' by pugilists in order to toughen himself up, and after vigorous outings on the Thames in what are known as tub pairs and tub fours, used to bathe, winter and summer, in the now-polluted waters of the river Wandle where it entered the Thames at Wandsworth, before setting off to work in 'The Drapery'. He was a rowing man before everything, even before his business. So great was his passion for rowing that he had left his newly married wife (my mother-to-be) at the wedding reception at Pagani's in Great Portland Street on learning that it was just coming on to high water at Hammersmith and had gone down to the river by cab for what he described as 'a jolly good blow' in his double-sculler with his best man, who eventually became my godfather, returning hours later to his flat to find his bride in tears and having missed the boat train for Paris where the honeymoon was to be spent at the Lotti. His ambition was that I should win the Diamond Sculls at Henley, and in this ambition he was aided and abetted by my godfather, a crusty old Scot if ever there was one, who had himself won the Diamonds and the Stockholm Olympics in 1912.

To help me to victory in this and life's race my father insisted that my bowels should open at precisely the same moment every morning (this was at a time when certain Harley Street surgeons were advocating the removal of whole stretches of their patients' digestive tracts in the belief that whatever was passing through would emerge at the other end with as little delay as possible and thus avoid 'poisoning' the owner). In addition, he made me sniff up salt and water so that my nasal passages might remain equally clear, and have a cold bath each morning, winter and summer. When I was older I learned from him that besides keeping one in trim, cold baths were an aid against filthy thoughts, although I never found them to be of any remote use for this purpose (as useless as telling an Eskimo that he won't have filthy thoughts if he sits on an iceberg). In the early mornings I accompanied him on brisk trots along the towing path at Hammersmith, or down deserted suburban streets, punting a football, which I thoroughly enjoyed. At the age of six or so I learned to row our sumptuous, *Three Men in a Boat*-type, double-sculling skiff, which was kept at Richmond and in which we used to go camping 'up-river', wielding one enormous scull as an oar. In the same way my mother, who had been a model girl in my father's firm, and who was more than twenty years younger than he was and still went with him to Paris long after they were married to buy models from Poiret, Chéruit, Patou and others which were made to her lath-like proportions, had been turned into a very stylish oarswoman.

Although my mother no longer worked for Harrods she had not lost her enthusiasm for the store. She was no mean spender, my mother, and she went through the place like a combine harvester on my behalf. This trait of extravagance was belied by her rather sad, tranquil expression when in repose, just as it belied her vivacity and fondness for company.

Thus a complete set of gear awaited me on 6 December when I turned up, most of it procured from Harrods 'on account'. It included the pram with its fringed sun awning, an 'extra' bought in anticipation that I would survive until the summer of 1920, the 'French bassinet' with its iron stand and an arm which supported the *baldacchino* of fine cotton voile under which I lay tippling gripewater; a white-enamelled folding-bath, complete with soap dish containing a cake of Harrods's 'own make' baby soap and a sponge tray with one of their 'specially selected sponges' in it; a spring balance with a wicker basket, capable of weighing babies up to twenty-five pounds, which was later converted for use in the kitchen by the substitution of a metal pan for the basket; and a nursery screen. Surrounded with this and other equipment (I cannot remember the lot, but this is some of what survived until I was older and could remember), I must have looked like a beleaguered traveller behind a makeshift breastwork awaiting a charge by fuzzy-wuzzies.

If anything ran short which she thought was better ordered from Harrods than bought locally, or she saw something that caught her fancy in their catalogue, my mother used to say, 'I'll get on the telephone to Harrods,' the telephone being a solid, upright metal instrument with a separate receiver, weighing pounds, which householders were beginning to find useful for laying out the first wave of post-war housebreakers who were now just beginning to come back into circulation, a process that could operate in reverse if the burglar picked it up first. To my mother, the possibility of being able to telephone for a consignment of Harrods's Finest French Sardines in Olive Oil or some bottles of Rubinat Water, which she used as an aperient, and receive them that same afternoon, delivered in a shiny green van with the royal arms on it, was magic.

What was probably my first visit to Harrods, the first I can remember, anyway, took place on the occasion of the rigging out of Lily in Nurses' Uniforms, at that time on the first floor. I remember it not because it was intrinsically interesting but because it took ages and because at one stage all three of us, together with a saleswoman, were crammed into a very small, stifling fitting-room, like the Marx Brothers in the cabin scene on the transatlantic liner in *A Night at the Opera*.

From Nurses' Uniforms I was escorted to Children's Hairdressing, also on the first floor. There, the infant Newby was shaped up again, after having spent some happy minutes snipping away at his noddle with a pair of stealthily acquired nail scissors while seated incommunicado on the pot in front of the gas fire which by this time had replaced the coal fire in my nursery at Three Ther Mansions.

To me Harrods was not a shop. It was, apart from being the place where I had my hair cut, a whole fascinating world, entirely separate from the one that I normally inhabited. It was a world that, although finite in its extent (it covered thirteen acres), I never explored completely, never could, because although at the early age of which I am writing I did not realize this, it was one in which fresh vistas were constantly being revealed, as the management either opened up new, sometimes ephemeral departments or introduced innovations within existing ones.

For instance, in 1929, following Lindberg's solo crossing of the Atlantic, they opened up an Aviation Department and taught some of their customers to fly. Eventually, when there were not enough potential aviators left untaught among their customers to make it worthwhile keeping it open, it quietly faded away.

'Hold my hand tight, or you'll get lost,' my mother used to say, as she moved through the store, browsing here and there like some elegant ruminant, a gazelle perhaps, or else walking more purposefully if she was on her way to some specific destination, as she often was. My mother was not the sort of person who only entered Harrods in order to shelter from the rain. Once she was in it, she was there as a potential buyer.

And I did hold tight. Get lost in Harrods and you had every chance, I believed, in ending up in the equivalent of that undiscovered country from whose bourne no traveller returns, which when I became a grown-up with an account of my own I located somewhere between Adjustments and Personal Credit (which comprehended Overdue Accounts) and the Funeral Department, for those whose shopping days were done but whose credit was still good, both of which were on the fourth floor.

This world, which I was forced to regard from what was practically floor level, was made up of the equivalents of jungles, savannas, mountains, arctic wastes and even deserts. All that was lacking were seas and lakes and rivers, although at one time I distinctly remember there being some kind of fountain. The jungles were the lavish displays of silk and chiffon printed with exotic fruits and lush vegetation in which I was swallowed up as soon as I entered Piece Goods, on the ground floor, which made the real Flower Department seem slightly meagre by contrast. The biggest mountains were in the Food Halls, also on the ground floor, where towering ranges and isolated stacks of the stuff rose high above me, composed of farmhouse Cheddars, Stiltons, *foie gras* in earthenware pots, tins of biscuits, something like thirty varieties of tea and at Christmas boxes of crackers with wonderful fillings (musical instruments that really worked, for instance), ten-pound puddings made with ale and rum and done up in white cloths, which retailed at £1.07½ ($4.17) the month that I was born. Some of these apparently stable *massifs* were more stable than others and I once saw and heard with indescribable delight a whole display of tins of Scotch shortbread avalanche to the ground, making a most satisfactory noise.

In the great vaulted hall, decorated with medieval scenes of the chase, and with metal racks for hanging the trophies of it, where Harrods's Fishmongers and Purveyors of Game and the assembled Butchers confronted one another across the central aisle, there were other mountainous displays of crabs, scallops,

Aberdeen smokies, turbot and halibut, Surrey fowls and game in season on one side; and on the other, hecatombs of Angus Beef, South Down Lamb and Mutton.

The savannas were on the second floor, in Model Gowns, Model Coats and Model Costumes, endless expanses of carpet with here and there a solitary creation on a stand rising above it, like lone trees in a wilderness.

To me unutterably tedious were the unending, snowy-white wastes of the Linen Hall, coloured bed linen, coloured blankets, even coloured bath towels, except for the ends (headings) which were sometimes decorated with blue or red stripes, being – if not unknown – unthinkable at that time (coloured blankets, usually red, were for ambulances and hospitals). In it articles were on sale that not even my mother was tempted to buy: tablecloths eight yards long to fit tables that could seat two dozen guests, sheets and blankets ten feet wide, specially made to fit the big, old four-poster beds still apparently being slept in by some customers, in their moated granges.

Higher still, on the third floor, were what I regarded as the deserts of the Furniture Departments. It took something like ten minutes to get around these vast, and to me as un-interesting as the Linen Hall, expanses, in which the distances between the individual pieces were measured in yards rather than feet.

This 'Harrods's World' even had its own animal population in what the management called Livestock up on the second floor, what customers of my age group and most grown-ups called the zoo. In it the noise was deafening, what with macaws that could live for sixty years or so, Electus parrots in brilliant greens and reds and purples, according to sex, parrots that could speak – they had to pass a test to ensure that they did not use bad language – and other rare Asian birds, as well as puppies, kittens, guinea-pigs, mice, tortoises, armadillos and Malabar squirrels. I got my first mouse at Harrods.

But the greatest treat of all was a visit to the Book Department. I was not allowed to visit the Toy Department, except for my birthday, or at Christmas. In fact it was not very interesting except at Christmas time when it expanded for a month or two, then contracted again when the sales began in

January, until the following November or December; and it was never as good in those days as Hamleys Toy Shop in Regent Street.

Although my mother refused to supply me with toys on demand on these journeys through Harrods (for that is what they were to me), she would always allow me to choose a book. The first book I can ever remember having, a Dean's Rag Book, printed on untearable linen, came from Harrods, although even then I found it difficult to think of something printed on linen (or whatever it was) as a book. Once I was in the Book Department it was very difficult to dislodge me, and it was only because I was actually being bought a book that I left it without tears, and to this day I find it almost impossible even to walk through this department en route elsewhere, without buying a book I didn't know I wanted.

Beyond the Book Department was the huge, reverberating, rather dimly lit Piano Department, where salesmen who dressed and looked like bank managers used to hover among the instruments, trying to put a brave face on it when I ran my fingers along the keys of their Bechstein Grands as we passed through, probably on our way to Gramophone Records, of which my mother, who loved dance music and dancing, had already amassed a large collection. Sometimes a visitor to the store who was also a pianist would take his seat at one of the grand pianos and this otherwise rather gloomy room would be filled with wonderful sounds.

It was from this department that there emanated, by way of Accounts, a bill made out to my father for one of these grand pianos, at a cost of something like £125 ($531) but expressed in guineas, which when it was finally sorted out was reduced to one of about £1.25 ($5.31) for a couple of visits by a piano tuner to Three Ther Mansions in order to tune our modest, upright Chappell, the grand piano having been charged to him in error. Until long after the Second World War, really until they installed a computer, Accounts had a dottiness about them that was sometimes, but not always, endearing; and until the computer was installed it was perfectly possible to order a pound or two of smoked salmon to be delivered from the Food Hall and not actually pay for it until three or more months later.

After seeing some or all of all this, for if my mother went to Harrods in the morning she would also spend part of the afternoon there, she would whisk me off to the Ladies' Retiring Room on the fourth floor where she freshened us both up before taking me to lunch in the Restaurant where, jacked up in a special infant's chair which elevated my nose and mouth above what would have been, sitting in an ordinary chair, the level of the table, I ate what at that time was my favourite meal, half portions of tomato soup, fried plaice and creamed potatoes. After this we again repaired to the Ladies' Retiring Room, which I recall as being rather grand and commodious, for a brief period of doing nothing.

It is now many years since I have visited this room. Even when I used to visit it fairly frequently you had to be pretty young to be allowed in if you were a man; but I distinctly remembered on one occasion seeing what even I could recognize to be a very elegant, very emaciated lady who was wearing a *bandeau* on her head, which was very fashionable then – or could it have been an ice-pack? – and who was reclining on a wicker chaise longue and uttering a series of 'Oh God's' at intervals.

At least I thought I could remember. However, when I reminded my mother of this incident, in the mid-1960s, she said it could not possibly have been at Harrods.

'No, it wasn't Harrods,' she said, 'it was Dickins & Jones. I remember they were building Liberty's, the half-timbered part that looks like an old house, in Great Marlborough Street, almost opposite Dickins & Jones. It must have been 1923. The builders were making a terrible noise with drills and things, and that poor girl, the one you remember, she was very smart, had a terrible headache. She'd probably been to a party the night before. There were lots of parties then. Besides, there was never anywhere to lie down at Harrods so far as I can remember, except in the Furniture Department, and that would probably have meant buying a bed or a sofa.'

Although my mother's recollection of where this event actually took place also made me an honorary member of Dickins & Jones's Ladies' Retiring Room as well as Harrods's, I never really liked the Regent Street store. Partly because in my opinion there was not much worth looking at, although I liked the smells in Scent – no zoo, no Book Department. But my real reason for disliking it was because my mother used to take me up to one of the Fashion Departments and display me to the buyer, whom she knew, and to the salesgirls, just as she used to do at Harvey Nichols, Debenham and Freebody, and Marshal & Snelgrove, a process which to me seemed to take an eternity.

After this mandatory rest in Harrods's Retiring Room, my mother used to take me to the Picture Gallery, which I loved and still do. Then, as until recently, the strictly representational nature of the pictures on view underlined as nothing else does in the store, except perhaps in Gifts, the basically unchanging taste of Harrods customers.

In it hung paintings, most of them in cheerful colours: of clipper ships sailing up the Channel under stunsails; the pyramids with fork-bearded, armed nomads and their camels silhouetted against the sunset; lovers in gondolas passing beneath the Bridge of Sighs; bewigged eighteenth-century gentlemen dallying with ladies in perfumed, English rose gardens; scantily but always decently clad Circassians languishing under the wild eyes of prospective buyers in Moorish slave markets; snow-covered Alpine and Rocky Mountain peaks, bathed in shrimp-pink evening light; unlubricious nudes; ducks flighting in Norfolk; Highland stags at bay; Indian tigers and herds of African elephants sufficiently hostile-looking to make it pretty certain that the artist had painted them from photographs, or while up a tree; race horses at Newmarket, and all the animals too large to be stocked in Livestock; riots of cardinals surprising clutches of nuns or, surrounded by empty jeroboams, complimenting the chef on an unusually rich dinner in some French *palais*. Here, a world beyond Harrods's world opened out before me.

Here, in Harrods to this day I can evoke the happiness and more occasionally the miseries of the first twenty-five years or so of my life. It was where I went in Harrods, rather than what I bought or what was bought for me, that I remember, the *genus loci* of the place: which is no doubt what the now long-forgotten architects Stevens and Hunt (the latter of whom was immortalized as Munt by Sir Nikolaus Pevsner in his great work, *The Buildings of England*) intended: the lonely staircases, which no one ever used because everyone travelled by lift, led down so quietly that standing on them you could hear the wind whining round the building; the ceilings supported by green marble columns decorated with gilded Egyptian motifs on the ground floor in Gifts, and the elaborate white plaster ceilings on the first floor, rather like the decoration on what they called 'Our Own Wedding Cakes' which, like their own Christmas puddings, bread, pastries, sweets, chocolates, veal and ham pies, and goodness knows what else, were made in their own factory over the road. And there were the, to me, beautiful bronze lifts, embellished with what looked like strips of woven metal, one of which still survives. One of these lifts in those years after the First World War was operated by an ex-serviceman with one arm, just conceivably the 'Ex-Service Man. Loss of right arm, seeks situation as Window Dresser or Shopwalker', who advertised in *The Times* on the day I was born but failed to secure one of these positions.

Also on the ground floor, not far from Gifts, was a vast room which housed Jewellery, Silver, Optical and Cutlery. Jewellery was furnished with little, green leather-topped tables, with looking-glasses and red-shaded lamps on them. At these tables what appeared to me, over a period of thirty years or so, to be the same salesmen sat opposite their customers, breathing discreetly at the conclusion of a deal (all that changed was the price and that only latterly), 'Thank you, sir/madam, an excellent choice, if you will allow me to say so. That will be two thousand, two hundred and twenty-five guineas (£2336.25 or $9929). May I ask if we have the pleasure of having an account with you at the store? Will you, uhum, be wishing to take the necklace with you?'

My first wristwatch, the first really adult present I ever received, made specially both for schoolboys, and for Harrods, came from this department, in the autumn of 1927. It was to encourage me 'to be a man', as my father put it, when I went to my prep school, Colet Court, a prospect which at that time, not yet being eight years old, I found terrifying; but nothing like as awful as the reality. In Cutlery, besides canteens of silver and electro-plate in oak and mahogany cabinets, they carried stocks

43

of fighting and hunting knives, ready for travellers who needed to give the *coup de grâce* to dying tribesmen or wounded bears in the Balkans. Until long after the Second World War they had show cases filled with regimental swords all ready, apart from being sharpened, for the next great struggle.

Next door to Jewellery, Silver, Optical and Cutlery, in a kind of limbo between it and Gentlemen's Outfitting (now the Man's Shop) was the Boys' Shop. In it they sold all the gear you needed to be 'privately educated' in Britain, at preparatory and what are so oddly known as public schools, with the names of more respectable ones emblazoned on the oaken fixtures. In this department over the years I was successively fitted out with flannel shirts, flannel shorts supported by belts striped in the school colours with snake-head buckles, floppy grey flannel sun hats, navy sweaters with collars emblazoned with the school badge, blazers, white trousers, straw hats, black jackets, striped trousers, starched white collars with round bottoms which showed off nicely the brass collar stud, and, almost unbelievably, bowler hats.

To this day passing the site of what was once this department which was linked with the Man's Shop by flights of symbolic steps and an equally symbolic tunnel, I still experience the feeling of doom that descended on me like a pall during the last ten days or so of the summer holidays, a feeling aggravated by Harrods with their triumphant slogan, constantly reiterated in their catalogues and window displays, 'Back to School!' How I hated them. It is not surprising that for years one of the difficulties (which the management admitted) in getting grown-up customers to patronize their ample and sumptuous Man's Shop was that many of them had never recovered from their traumatic experiences in the Boys' Shop.

Nevertheless, when I returned to England from a prisoner-of-war camp in Germany in the spring of 1945, I spent part of my first traumatic morning of freedom in it trying, successfully it turned out, to buy a pair of corduroy trousers without the obligatory clothing coupons with which I had been issued but which I had already succeeded in temporarily mislaying.

'Dear, dear, sir,' said the very elderly salesman when I explained my predicament, eyeing the enormous 'Battledress anti-gas' with which I had been issued, presumably in error, in Brussels, after my liberation. 'We can't have one of our old customers without a change of trousers, can we? That would never do. Mum's the word, but here in Harrods we've got more gentlemen's trousers than there are coupons in the whole of the United Kingdom.'

5

Westward Ho!
(1925)

And not by eastern windows only,
   When daylight comes, comes in the light,
In front the sun climbs slow, how slowly,
But westward, look, the land is bright.
A. H. Clough, Say Not the Struggle
Naught Availeth

In 1925, when I was five and a half, we embarked on what, so far as I was concerned, was the most ambitious holiday I had ever had. In summer my father took a cottage at Branscombe, at that time a very rural and comparatively unvisited village in South Devon, between Seaton and Sidmouth. It promised to be a particularly exciting time as my father had decided that we should travel there from Barnes by motor. This meant that most of our luggage had to be sent in advance by train from Waterloo to Honiton, a market town on the main line to Exeter; at Honiton it was picked up by a carrier and transported the ten miles or so to Branscombe by horse and cart. Others taking part in this holiday, although they did not travel with us, being already foregathered there, were my Auntie May (the aunt who had accompanied my mother and me on the memorable visit to Godshill) and her husband, Uncle Reg. Before the war Uncle Reg had worked as a journalist in Dover on the local paper and in this capacity had been present in 1909 when Blériot landed on the cliffs, having flown the Channel. During the war he had been in the navy in some department connected with propaganda. Later he became editor of the *Gaumont British Film News*. He was very urbane and elegant. He was later on good terms with the Prince of Wales for whom he used to arrange film shows at Fort Belvedere, and for the Royal Family at Balmoral. For these services he was presented with cufflinks and cigarette cases from Plantin, the court jeweller, as well as other mementoes. He preferred to be called Reginald rather than Reg, but no one ever did so. They put up in the village pub where we, too, were to take our meals.

The third party was made up of three fashion buyers for London stores, Beryl, Mercia and Mimi Bamford, all of whom were friends of my mother, particularly Mimi, and their mother. All three were unimaginably elegant, often almost identically dressed in long, clinging jerseys and strings of amber beads, and they were surrounded by what seemed to be hordes of extremely grumpy pekinese who did not take kindly to the country. Their mother, who did not take kindly to the country either, was even more formidable. She owned a Boston Bulldog called Bogey, which had had its ears clipped, a practice by then declared illegal. Like her daughters she was immensely tall, and must at one time have been as personable as her daughters, but even I could recognize that she was incredibly tough, if not common.

'She didn't ought to 'ave 'ad 'im,' was the comment she made about me, by now a boisterous, active little boy, to my Auntie May while we were at Branscombe, 'she' being my mother; an anecdote that my aunt eventually told me, which she did with an excellent imitation of the old lady's gravelly voice, having put off doing so until only a few years before her own death in 1974 in order, as she put it, to spare my feelings.

Neither Beryl nor Mercia nor their mother ever went to the beach, or even set eyes on the sea, the whole time they were at Branscombe. For Beryl and Mercia the seaside was Deauville. What Branscombe was to them is difficult to imagine, or they to the inhabitants. Only Mimi relished the simple life.

The morning of our departure from Three Ther Mansions was a fine one. We were seen off by the head porter of the flats – gratuity – and by Ellen, the cook/housekeeper, who had taken Mrs George's place in a resident capacity. At that time Ellen was probably in her late forties. She had smooth black hair parted down the middle with some white strands in it, a pale face and a rather forbidding, if not sinister, appearance – perhaps secretive is more appropriate; in retrospect I think what she most resembled was a female poisoner – and she was stiff, and starchy, or at least her aprons were starchy. In spite of her apparent grimness or strangeness or secretiveness Ellen was always very kind to me, especially when my parents were away, and I think that when they returned she resented their presence.

I found Ellen disturbing in a way which I could not have explained to her or to anyone else, not even to myself. Sometimes I used to ask her to take me in her lap and cuddle me. If she did, and sometimes I would have to ask her several times before she agreed, she would fondle me in a detached, offhand way that made me all the more determined to make her love me, although I did not really love her; but I was never successful.

Whether it was the Napier with the gleaming brass-framed windscreen, or the much more modest Citroën which succeeded it at about this time, whatever we were travelling in was a pretty close fit for the five of us, even with the picnic baskets and our overnight suitcases strapped on the carrier at the back. (I can't remember which it was and anyone else who would know, or care, is dead.)

There was Mr Lewington, the chauffeur from around the corner in Fanny Road, who was driving us down to Devonshire and then bringing the car back to London – all attempts by my father and mother to learn to drive themselves had been attended by what might quite easily have been fatal results (my father had demolished the façade of a garage; my mother had left the road on Barnes Common and travelled some distance overland before coming to a halt). There was my father who sat next to him, from which position he was better able to keep an eye on the behaviour of other road users, pedestrian or otherwise, and if necessary stand up and rebuke them for some actual or imagined infringement of the rules of the road; and in the back, besides myself, there was my mother and Kathleen, usually known as Kathy, a sweet girl of fifteen or so with long auburn hair down to her waist who had replaced Lily and now 'helped out' with such chores as taking me to school. Kathy wore ordinary clothes. My mother had given up any high-falutin' ideas about uniforms when Lily left and anyway I was now much too old to have a real nurse. Kathy was as excited about the trip as I was, never having been out of London before. There we sat with a travelling rug over our knees made from what looked like a leopard skin but was really a very costly length of woollen material from Paris, bought by my father with the intention of making from it a model coat, but pinched by my mother who had it made into a rug. In front of us was a second windscreen, a sheet of metal-framed glass, that would be as lethal if shattered in an accident as the front one would be, that could be folded down to form a picnic table when the machine was at rest. There was no such thing as safety glass in common use. In winter it was jolly cold in our open motors, even with the 'lid' up, and then everyone except the driver and me, because my legs were too short, was provided with what were known as Glastonbury Muffs, huge boots lined with sheepskin, each holding two feet. If it was really cold these boots had a sort of pocket in them which could be used as a receptacle for a hot water bottle; but I do not believe we ever went motoring with hot water bottles.

Branscombe was just over a hundred and fifty miles from Barnes by the direct route. Usually, when travelling with my father, we did not follow the direct route, he being as curious about what lay on either side of the direct route to anywhere as I myself was to be, years later. At Staines, we got down to admire the river and talk to a waterman of my father's acquaintance at one of the boathouses. (He knew every waterman of any consequence between Putney and Henley.) At a place called Virginia Water we visited some exciting ruins brought all the way from Leptis Magna in Tripoli, and an even more exciting waterfall full of enormous rocks.

Then we drove on between miles of rhododendrons and across commons (over which, years later, in 1940, I would crawl on all fours armed to the teeth with 'token' wooden weapons because all the real ones had been taken away to give to the 'real' army after Dunkirk), my father with his quarter-inch to the mile Ordnance Survey 'Touring' Map at the ready to deal with any navigational problems and, as a member of the Royal Automobile Club and the Automobile Association, with both badges on the front of his motor car (the RAC's was grander), returning the salutes of the patrolmen of both organizations, (according to whose area we were passing through), men with wind-battered faces, wearing breeches and leather gaiters, who were either mounted on motor-cycle combinations or, in some more rural parts, on pedal cycles. If they failed to salute, members were advised to stop them and ask the reason why.

Among the interesting things we saw on this early June morning was a nasty accident, with splintered windscreens and lots of blood, which, although Lewington slowed down so that we could have a better view, I was not allowed to see properly, my head being turned the other way by Kathy, much to my disappointment. We also saw traction engines and enormous machines called Foden lorries, fuelled with coal and belching steam and smoke and red hot cinders; and a 'police trap', set up by a couple of constables armed with stop-watches, one of whom had emerged from a hedgerow in which he had been lurking to flag down a luckless motorist for 'speeding'. Until November 1931, the speed limit in Britain for all motor vehicles was officially twenty miles an hour. And we saw lots of tramps, most of them, like us, heading west, one of them a diminutive elderly man wearing a bowler hat and a wing collar, who was helping his equally elderly wife to push a very old-fashioned, high-wheeled perambulator with all their possessions in it along the road. And once we saw a team of huge English carthorses pulling a wagon with an enormous tree trunk on it.

At one point we came to a magic place where a road forked away from the one on which we were travelling, a place that I never forgot and was therefore subsequently able to identify, although it was years later. There, sheltered by trees was a grass-grown open space, with a number of low, barn-like buildings disposed about it. Some had slate roofs; some were thatched; some of the more important-looking ones – that might have been part of a farm – were built of cob, a mixture of clay and straw, and had enclosing walls of the same material, which were also thatched or tiled, as they had to be, otherwise they would have melted away in the rain, something I had never seen before.

And it was here, at this moment, as if to set a seal on my memory of it, a memory that would endure for the rest of my life, and embody so many feelings that I could never express, that my father half stood up in the front of the car and shouted over the top of our windscreen, 'Hilda! Look at those buildings! How they're built! That means we're in the West.' And he was right. For this was Weyhill, the Werdon Priors of Hardy's *Mayor of*

Casterbridge, famous for its ancient six days' fair beginning on old Michaelmas Eve (10 October), one of the most ancient in Britain, to which a line in *The Vision of Piers Ploughman*, c.1360, 'To Wy and Wynchestre I went to the fayre', perhaps refers. Horses, sheep, cheese and hops were the principal things sold at the fair and as many as 150,000 sheep once changed hands here in a day. The second day of the fair (old Michaelmas Day) was the great hiring-day for farm servants and labourers in this part of Hampshire and the adjoining districts of Wiltshire, the carters appearing with a piece of plaited whipcord fastened in their hats, the shepherds with a lock of wool. For many years now I have watched the at first gradual and later the accelerated decay of these strangely beautiful buildings and, recently, I have witnessed the final destruction of the best of them.

We continued our journey across Salisbury Plain. There were sheep on every horizon now, and from time to time we passed clumps of ugly buildings. (My mother said they were military buildings left over from the war.)

There were more sheep around Stonehenge and I remember hearing the skylarks overhead, so high that I could barely see them, and running my hands over the surfaces of stones so huge that I could scarcely apprehend them. The only other visitor was a vicar in a dog-collar who had driven there with a pony and trap, and I remember thinking how much funnier we looked with our smart clothes and our shiny motor car in such a place than he did with his old suit and his rather yellow dog-collar, and his pony and trap.

A little further on we turned off the main road and followed a track that led past a number of grassy mounds, and there we had our picnic, not in the motor car using the collapsed windscreen as a picnic table but as we always did, unless it was raining, on rugs on the grass.

It was a memorable picnic, even though picnics arranged by my mother and father were always memorable. It is not just family pride that makes me say so. They really had a flair for picnics. Everyone said so.

There was a big pie, whether it was veal and ham with eggs embedded in it (not extruded through it as they are today) or pork is of no importance. I only know that like every other pie at every other picnic for which my father provided pies it had a design embossed on the crust (acorns and thistles were popular), and was the most delicious sort of juicy pie imaginable. And there was a ham, which my father had bought in case the meals at the pub at Branscombe were not always up to scratch, and an ox tongue and Stilton cheese, and lettuce, tomatoes and spring onions, and loaves of bread like cannon balls and Huntley and Palmer's Oval Water Biscuits, and home-made mayonnaise, and Ventachellum's Sweet Sliced Mango Chutney, and fresh fruit salad, and to drink there was Whiteway's Dry Devonshire Cider for the grown-ups, and for me lemonade, made at home by Ellen. After which everyone except myself had 'forty winks'.

Meanwhile I climbed one of the mounds, and looked out over an endless, undulating stretch of grass that looked like a heaving sea and reminded me, perhaps because I had eaten too much, of the real sea, also under a cloudless sky, the time we had gone to Sark by ship and my mother and I had been dreadfully sick, partly because of the heavy swell that was running, partly because we had eaten something called haricot mutton in the dining-saloon. Then, below deck, we had been ministered to by terrible-looking women dressed in black, called stewardesses, who handed us earthenware vessels a bit like chamber pots to throw up in, and who called me 'ducks'.

While we were having the picnic, in answer to my question my father told me that the mounds, on one of which I was now standing, were really tombs and there were dead people under them and probably treasure, too. And as the one on which I was standing was already eroded on one side so that the chalk showed through I got a bit of stick and tried to dig my way into it with the intention of looking at one of my ancestors, and perhaps finding a crock of gold. However, in spite of these incentives, I soon gave up and simply sat on top of the tumulus, listening to the larks' song rising and falling, looking out at the sheep lying close together, like white rugs on the green grass, until it was time to take to the road again.

On to Salisbury (one of my father's detours), where we visited the cathedral, and we were taken over it by an extraordinarily enthusiastic white-haired guide. At one moment we were up under the roof with him among enormous timbers that made it seem like a forest of leaning trees; the next he was lining us up, numbering us off from right to left, and teaching us bell ringing with handbells in the vestry.

We had tea in Dorchester, then drove on into the eye of the now declining sun, with occasional entrancing views, ahead and away to the left, of the shimmering sea, in Lyme Bay. At Morecombelake we visited what looked to me like a very old factory in which local women were rolling out what looked like huge musket balls of dough on wooden tables. Put in the oven they emerged later as what are called Dorset Knobs; and my mother bought a large tin with a lovely red, white and blue label on it ornamented with towers and castles.

Not long afterwards we descended a very steep and winding road with a notice reading 'Engage first gear' at the top of it, and entered a pretty little seaside town at the mouth of a deep valley. This was Lyme Regis and here we were to spend the night at an inn where my father had taken rooms. Why he had chosen to do so when Branscombe was only about twelve miles away was a bit of a mystery. He may have thought we would arrive much later, allowing for punctures and mechanical breakdowns. Whatever the reason it was quite an expensive decision, but it might have been worse. I was half price and did not qualify for dinner, having eaten a large tea. I would be content with a hot drink and some Bovril sandwiches. And Lewington and Kathy would both qualify for the special terms offered for chauffeurs and servants, which included supper, bed and breakfast. Only my parents paid the full rate for their double room, dinner and breakfast.

But our arrival at Lyme Regis did not mean that it was the end of the day for me. Before it was time for bed I had already 'done' Lyme Regis. I had seen the studded door of the old lock-up with its inadequate airholes, now incorporated in a picturesque but more modern building; I had visited a museum full of fossils and what called itself a 'fossil depot', where one could actually buy them, except that they were very expensive. Nevertheless, in the course of the evening, thanks to Lewington, I acquired a collection of my own. Walking along the beach after the abortive visit to the fossil depot and climbing over a number of groynes, we eventually came to some dark and crumbling cliffs. By this time of the evening they were already in shadow and made a striking contrast with the enfilade of great sandstone cliffs which stretched away beyond them towards Portland (one of them the highest of its kind in southern England), the summits of which were glowing like golden castles in the evening sunlight and would continue to do so for some time after it had departed from the remainder of the landscape. That evening, by some strange chance, perhaps because there had recently been a heavy rainstorm, fossil ammonites could be seen protruding from the stiff clay of the cliff, and by dint of some energetic scrambling, prodding with long bits of driftwood, and some stone throwing Lewington, who was young and active, managed to dislodge several fine specimens, making his blue serge chauffeur's suit rather muddy in the process. Altogether, apart from not having been allowed to bathe when we reached Lyme Regis, from my point of view it had been a thoroughly successful day.*

(The inn at which we stayed that night, or another very similar inn, was the scene, more than thirty years later, of a disgraceful incident in which I was myself involved. Not long after the Second World War, while on a walking holiday with a great friend along the coasts of Dorset and Devon in the depths of winter, the two of us arrived at Lyme Regis where we put up at one or other of the inns. It happened to be New Year's Eve and, in the course of the evening, having fulfilled a lunatic

---

*It was in these cliffs at Black Ven, or Vein, on a coast that is the epitome of Victorian romanticism in the West, that Mary Anning, then aged twelve, the daughter of a vendor of curiosities, known as the *Curi-man*, in 1811 discovered in the Lower Lias (the lowest series of rocks of the Jurassic system) the first known skeleton of an ichthyosaurus, a marine reptile twenty-one feet long with a porpoise-like body, dorsal and tail fins and paddle-like limbs of the Mesozoic era that began 225,000,000 years ago and lasted for about 155,000,000 years. It took ten years to remove it from the cliff and it was acquired by the British Museum for the sum of £23. Mary Anning subsequently discovered the first known specimen of the plesiosaurus, a creature with a long neck, short tail and paddle-like limbs, and of the pterodactyl, a flying reptile having membranous wings supported by an elongated fourth digit and, if reconstructions are anything to go by, of a terrifying aspect. A painted window in her memory was put up in the church at Lyme Regis, partly subscribed for by the Geological Society of London who described her in an obituary address as 'the handmaid of geological science' on her untimely death at the age of forty-seven.

ambition to visit every pub in the town, and previously weakened by having walked some twenty-five miles, we became stupendously drunk. It was as a result of this that in the early hours of the morning, being incapable of finding the light switches, we both peed down the bend in the staircase from an upper floor on to what I recall to have been the visitors' book on the reception desk, under the impression that we were in some sort of lavatory. How we escaped detection is a miracle and a mystery only explicable by the fact that the staff of the hotel must have been revelling too.)

The next morning when I looked out of the window, Lyme Regis was more or less obliterated, filled with a chill mist that was funnelling up into it from the sea, and in spite of various local prognostications it failed to disperse before we set off.

Sitting in the back, in spite of having our own windscreen and the travelling rug, we were glad of one another's proximity as the car groaned in bottom gear up the hill which seemed almost as steep as the one by which we had entered the previous evening, past bow-fronted buildings and more classical edifices. As we climbed above the town I had fleeting glimpses of rustic houses, partially hidden by flint walls and foliage, some of them with pretty verandahs. One house, which had a steep pitched, thatched roof that came down so low that it had to be supported by posts, looked as if it was sheltering under an over-size umbrella, and was so minute that it seemed impossible that any grown-up could enter it, let alone live in it; this was Umbrella Cottage, Lyme Regis, which still stands.

Then we left Dorset for Devon (a sign told us so), crawling along in high, open country with the headlights on, alone in the sea mist, as if we were the only living creatures in the world, apart from an occasional cluster of cattle at a hedgerow gate, or a couple of horses, their breath steaming; seeing not much else but an occasional tree, the telegraph poles as they came looming up one after the other through the mist, or the vague outline of some building, the mist clouding the windscreen so that several times Lewington had to stop, get down and clean the glass. It was mysterious and exciting. It was the kind of weather, I heard my mother tell Kathy, that one expected in 'the West'.

We descended steeply to the River Axe, crossed it by a narrow bridge, then waited for some time at a level-crossing for what looked like a toy train when it finally appeared to emerge from the mist and chug across our path, whistling mournfully, before being swallowed up once more; then climbed to another upland where, if anything, the mist was thicker, my father busy with the map now, until suddenly he told Lewington to turn into a lane on the left that was almost invisible in such conditions.

At first it ran between hedges through the same flat, upland country, then after a little while it began to descend and all at once we were out of the mist and before us was an enchanting, arcadian prospect, so enchanting that my father ordered Lewington, who although perfectly happy to stop for a motor accident was singularly obtuse when it came to such phenomena as views, to stop, so that we might admire it further.

There was no more wind and the mist was now above us on the high ground. Far below in the bottom of a narrow, sheltered valley and in the mouths of a couple of lesser combes which contributed to it, was a long, straggling village of thatched houses from the chimneys of which smoke was rising in tall, unruffled columns. The steep sides of the valley were decked with woods that seemed to hang above it and from the trees came the sound of a thousand disputing rooks. Even as we looked the mist began to dispel from the high tops, revealing long ridges running inland.

By the time we reached the village square, if the conjunction of three small roads at the point where a general shop and a butcher's stood could be dignified with such a description, the sun was shining from a clear sky. We had arrived at Branscombe and it was going to be a lovely day.

6

A Walk in the Sun
(1925)

Behind the Mason's Arms, the pub which stood next door to the cottage my father had taken for the summer and of which it formed a part, there was a yard surrounded by various dilapidated outbuildings and a piece of ground overgrown with grass and nettles which concealed various interesting pieces of rusted, outmoded machinery, the most important of which was an old motor car smelling of decaying rubber and dirty engine oil. The stuffing of what was left of its buttoned leather upholstery was a home for a large family of mice. This yard was to be the scene of some of the more memorable games I played with my best friend in the village, Peter Hutchings, whose mother kept a grocery, confectionery and hardware shop on the corner opposite Mr Hayman, the butcher's. It was from Peter Hutchings, who was killed while serving as a soldier in the Second World War, and whose name is inscribed with the names of fourteen other village boys who died in the two great wars on the war memorial at the entrance to Branscombe churchyard, that I learned the broad local dialect which was so broad that by the end of that first summer at Branscombe no one except a local inhabitant could understand what I was saying. 'Sweatin' like a bull 'er be,' was how Peter Hutchings described to me one day the state of his sister, Betty, confined to bed with a temperature, and it was in this form that I passed on this important piece of news to my parents.

There in the inn yard, in the long summer evenings, we used to sit in the old motor car, either myself or Peter at the wheel, taking it in turn, the driver making BRRR-ing noises, the one sitting next to him in the front making honking noises – the horn had long since ceased to be – as we roared round imaginary corners, narrowly missing imaginary vehicles coming in the opposite direction, driving through an imaginary world to an imaginary destination on an imaginary road, a pair of armchair travellers. In the back we used to put Betty Hutchings, if she was available, who wore a white beret, was placid, said nothing, apart from an occasional BRR, and was in fact an ideal back-seat passenger. Sometimes, if we felt like doing something 'rude', we used to stop the car and pee on the seats in the back, and Betty would pee too. This gave us a sense of power, at least I know it did to me, as I would not have dared to pee on the upholstery of a real motor car belonging to real people. Less courageous than my wife, who confessed to having peed on the back seat of a 'real' very expensive motor car stopped outside her parents house at her birthplace in the Carso, *and* smeared it with cow dung.

When we got tired of driving our car we ourselves used to become motor cars, tearing up and down the street outside making BRRR-ing noises of varying intensity as we changed gear, disturbing the elderly ladies who used to sit at their cottage doors making Honiton lace, pillow lace, appliqué and guipure, the principal manufacture of the village. Close by, over the hill at Beer where there were stone quarries, the quarry men's wives had made the lace for Queen Victoria's wedding dress in 1839, something that was still talked about in the neighbourhood more or less as if it had happened yesterday. It was these hideous BRRR-ing noises that no doubt prompted old Mrs Bamford, whose cottage also faced the main street, to utter the words, 'She didn't ought to 'ave 'ad 'im.'

But all this was in the future, that first day of our holiday.

The next morning I woke at what must have been an early hour and, obeying some mysterious summons, dressed myself in the clothes that I had worn the previous afternoon – white shirt, shorts, socks and white sun hat (I couldn't manage the tie unaided), brown lace-up shoes from Daniel Neal's in Kensington High Street (soon to be replaced by hobnail boots, bought for me at my earnest request so that I could be a real country boy, which took me ages to tie) and my pride and joy, a hideous red and green striped blazer with brass buttons that I had persuaded my mother to buy for me, much against her will, from Messrs Charles Baker, Outfitters, of King Street, Hammersmith, so that I should look more like what I described as 'a real schoolboy' rather than an infant member of the kindergarten at the Froebel School in Baron's Court. As school blazers were not made to fit persons as small as I was, when I was wearing it my short trousers were scarcely visible at all. Not even Messrs Baker appeared to know which school it was, if any, that had red and virulent green stripes as its colours. Then, having picked up a stout stick that I had acquired the previous day, I stole downstairs and let myself out into the village street which was deserted as it was Sunday morning.

At the side of the road, opposite what I was soon to know as Mrs Hutchings's shop, a little stream purled down from one of the side valleys, one of several such streams that, united, reach the sea at Branscombe Mouth; and there, under a brick arch, it issued from a pipe which supplied this lower end of the village with water, before burrowing under the road to reappear once more outside the shop. From here it ran away downhill over stones along the edge of a little lane with an old, ivy-clad wall on one side of it, chattering merrily to itself as it ran over the stones in a way that seemed almost human.

Here, in this narrow lane, the water had what looked to me like watercress growing in it, and it was so clear and delicious-looking that I got down and had a drink of it, only to find that it was not delicious at all and that it had a nasty smell. Later I discovered that Betty Hutchings used to drink from this crystal stream if she was not watched which was probably the reason why she sweated like a bull.

I continued to follow the stream, racing twigs down it, until it vanished into a sort of tunnel from which proceeded a delightful roaring sound. At the other end it emerged beyond a wicket gate to flow more placidly under a little bridge and in these calmer waters I spent some time stirring up the bottom with my stick and frightening some water beetles, the air about me filled with the droning of innumerable insects.

From this point it ran to join the main stream in the middle of the valley and here the path turned away from it to the left beyond a five-barred gate which, because I could not open it unaided, I squeezed underneath, to find myself in a beautiful and what seemed to me immense green meadow, hemmed in by hedgerows and huge trees and filled with buttercups, while high above it to the right were the hanging woods we had seen the previous morning from the car, the open down above them alive with gorse.

At the far end of this field the now augmented stream was spanned by a small wooden footbridge with a white painted handrail. When I eventually reached the stream, in spite of all these distractions and making a number of more or less unsuccessful attempts to spear on the end of my stick some of the older, harder sorts of cow flop in which the field abounded, and launch them into the air, the water tasted even funnier than it had done in the village outside Mrs Hutchings's shop and I stung myself on the nettles getting down to it.

Beyond the bridge the path continued uphill, dappled with sunlight under the trees, and here the air struck chill after the heat of the meadow. Then it dipped and suddenly I found myself out in the sun on the edge of an immense shingle beach which had some boats hauled out on it, and in my ears there was the roar of the sea as with every wave it displaced and replaced millions upon millions of pebbles. To the left it stretched to where a cluster of ivy-covered white pinnacles rising above a landslip marked the last chalk cliffs in southern England; to the right to the brilliant red cliffs around Sidmouth; and beyond it, out to sea, on what was a near horizon, for there was already a haze of heat out in Lyme Bay, I could see the slightly blurred outlines of what Harry Hansford, the local fisherman who lived opposite the blacksmith's shop, would soon teach me to identify as a Brixham trawler, ghosting along under full sail.

> 'Then felt I like some watcher of the skies
> When a new planet swims into his ken;
> Or like stout Cortez when with eagle eyes
> He star'd at the Pacific – and all his men
> Look'd at each other with a wild surmise –
> Silent, upon a peak in Darien.'

Many years were to pass before I read these words of Keats, but when I did the memory of that morning came flooding back.

There was a sound of feet slithering on the pebbles behind me. It was Kathy, panting slightly, as she had been running. 'Whatever did you do that for, Eric, you naughty little boy, without telling me?' she said. 'Your mum's ever so worried, and your dad, too. He'll be ever so cross if he finds out where you've been. You'd better keep quiet when you get back. I'll say I found you in the field.'

Together, hand in hand, we went back up the hill towards the village where the church bells were now beginning to announce the early service.

7

Journeys Through Darkest Hammersmith
(1928–36)

In the autumn of 1928 I was sent to Colet Court, the Junior School of St Paul's which stood opposite it in the Hammersmith Road. By a quirk similar (but not the same) to that which locates Harrods in SW1, when everything else in Knightsbridge is in SW3, Colet Court was in Hammersmith W6 and St Paul's in West Kensington W14.

As if to emphasize the connection between the two schools the architect or architects, who were obsessed with this material, had embellished both sets of buildings, which were constructed of purple brick, with what seemed like acres of shiny, orange terracotta. A large part of my schooldays were spent in these disturbing surroundings, which suggested some more restrictive uses than seats of learning, and it was quite common for boys waiting outside St Paul's in the evening for a bus to take them home to be approached by some stranger, often a foreigner, wishing to know what kind of 'institution' was this enormous pile of Victorian Gothic in the Hammersmith Road, and of what sort were the inmates. Eventually, after some exceptionally earnest or morbid sightseer, wishing to see and know more, had enquired at the porter's lodge, an order was promulgated to the effect that when such a question was posed by a member of the public we were forbidden to say that the place was a 'loony bin'.

A photograph prpbably taken in 1929, in the summer term of my first year at Colet Court, shows me, at least after the extensive retouching to which it had been subjected, to have been a healthy, cheerful-looking boy with large, protruding ears and a large mouth full of very slightly protruding teeth.

Because my ears stuck out (whichever ear I slept on flapped

62

over on itself, like an envelope) at night when I went to bed I was made to wear what were called ear-caps. They were made of cotton, had what looked like miniature fishing-nets on either side to keep the ears against the head, and were kept in place with a couple of bits of ribbon tied under the chin. Just as a hair-net or a bath cap imparts to the wearer an excessively senile or infantile appearance, according to whether the wearer is old or young, so did ear-caps, and for this reason I was extremely sensitive about wearing them, just as I had been sensitive, years before, about wearing red rubber waders on Bournemouth beach. These ear-caps proved to be useless for restraining ears as powerful as mine. After a few days my ears broke through the netting, and after I had worn out something like half a dozen pairs in a month, according to my mother, I was made to wear something more robust that looked like a pair of metal earphones with leather ear-pads in place of phones. However, these were so uncomfortable that I eventually refused to wear them at all, in spite of having been 'given the slipper' by my father to encourage me to do so. After this nothing more was done about my ears, as nothing was done about restraining my teeth, and eventually ears and teeth returned to where they should have been in the first place, of their own accord.

This photograph, my last studio portrait for many years, was taken by Mr Spencer, whose studio was in an early Victorian villa next door to St Paul's School, and it was he who had been responsible for photographing me ever since I first sat up, unaided, and wore nothing but a loincloth.

It was Mr Spencer – a mild, pleasant man who used to wax the ends of his moustache into needlesharp points with the aid of a preparation called *Pomade Hongroise*, an operation which he once performed for my benefit at my earnest request – who first engendered in me an interest in photography, although a number of years were to pass before I had the opportunity to gratify it. Mr Spencer's camera was a massive affair, made of mahogany and brass, which used glass plates. When he was going to operate it he used to put his head under a black velvet cloth and gaze into a ground-glass screen on which whatever he was photographing appeared upside down, which must have been disconcerting until he got used to it.

Mr Spencer had various backgrounds against which one could be photographed: woodland dells, palace balconies, simulated sunsets, that sort of thing. He could even paint in gnomes and fairies, and did so in one unforgettable picture of me, with a fringe and aged about four, wearing a pale blue knitted-silk round-necked pullover and shorts to match. Much to my disappointment, apart from the picture with the gnomes and fairies, which I do not think she herself could really have liked, my mother always insisted that whatever background I happened to be taken against should be eliminated, so that I invariably appeared in the finished portrait against a white or sepia nothingness.

Another important piece of equipment, with which Mr Spencer used to keep his younger sitters in good humour, was a little, brightly feathered bird, which spent the time in a small box when it was not called upon to play its part. Whether it was a real bird that had been stuffed, or an artificial bird, it is difficult to say; but the entire contraption came from France.

'Watch for the dicky bird!' Mr Spencer used to say, with his head under the velvet cloth, just before he was about to take a photograph, at the same time squeezing a red rubber bulb which caused the bird to pop out of the box and utter a few chirrups before disappearing.

Every morning in term time for my first two terms at Colet Court, I walked from Three Ther Mansions over the bridge and through the back streets of Hammersmith to school. On these journeys I was usually escorted by Kathy. My mother was often away now, travelling with my father. We did have a cook-housekeeper at this time called Mrs Hartland, who was large and puffed a lot. Mrs Hartland was more or less a facsimile of poor Mrs George, who had walked into the river by Hammersmith Bridge with her umbrella up. However, the walk to Colet Court, even if she came back by bus, would probably have done Mrs Hartland in. In the evening Kathy used to meet me and bring me home by bus.

All these rather complicated arrangements were necessary because my father insisted that I should walk to school each day, rain or shine, instead of being taken there on a bus from Ther Boiler, until I was considered old enough to travel by myself.

'The exercise will do him a power of good,' he used to say, as if I was some obese person who otherwise might spend the rest of the day with his feet up, instead of what I was, a rather skinny little boy who spent quite a large part of each day playing football, cricket, learning to box, training for sports day or else roaming around the playground with a friend pretending we were Sopwith Camels shooting up Fokker triplanes, doing our best, by keeping on the move as rapidly as possible, to avoid the gangs of bullies with which the place was infested.

'Breathe in deeply,' he used to say, 'when you're crossing the bridge.' And because this, too, would do me good, I did, inhaling the Thameside air and the nasty smells which came from a municipal rubbish tip and Manbré & Garton's saccharine factory which made the whole of Barnes stink when the wind was in the wrong quarter.

Once, that first winter, there was a pea-soup fog, a thicker version of the one in which, what now seemed long ago, we had returned to Barnes in the electric brougham from Pimlico, a sort that later, when I became interested in crime, reading the ghosted memoirs of ex-policemen from Boots Subscription Library, I associated with Jack the Ripper. In spite of its density I went to school just as if it had been any other day (something my mother would not have allowed if she had been at home, for such fogs were a menace to health, killing off innumerable people), despatched on this perilous journey by Mrs Hartland, who was much too much in awe of my father to countermand his orders.

Armed with a cap pistol for self-defence, holding Kathy's hand, something I would not have done in broad daylight in case some other boys from Colet Court saw me and pulled my leg about it, both of us wearing woollen mufflers over our mouths that – until they were washed – reeked of sulphurous soot, looking like a couple of robbers, we groped our way over the bridge and into Hammersmith in what had become overnight a void in which one could see nothing, except where here and there a gas lamp in the street produced a sickly yellow incandescence. In it we could hear the coughs and footsteps of other passers-by without seeing them until they were actually on top of us, the groaning of vehicles in low gear and the hoarse cries of men on foot armed with acetylene lamps who were trying to guide them through the murk. Eventually we arrived at school half an hour late, to find that those boys who had succeeded in getting there had already been sent home and, to my delight, repeated the whole adventurous process the other way round.

These narrow streets through which we made our way, now long since destroyed by wartime bombing or knocked down to make room for housing estates and flyovers, were where the poor lived. They even had the sort of names that, when I was older, I learned to recognize were reserved for the streets of the poor; because whoever was responsible for naming them, such as the official who named Fanny Road in Barnes, knew that it did not matter what sort of names they were given as the poor would never object to living in, for example, Distillery Lane W6.

In such streets endless rows of little two-storeyed terrace houses, built of fog-blackened London brick, stood back to back, each with its outside privy, separated by little yards in which the occupiers sometimes kept rabbits or carrier pigeons, or if they were large enough turned into little gardens; the sort of London houses which, if they have survived, have become something their builders and occupiers never dreamed of, desirable residences in streets with names that now have an equally desirable period flavour.

At one of these street corners there was a pub, taller or made to appear taller than the houses by a large sign with the name of the pub and the sorts of beer it sold inscribed on it in gold lettering, and curved to wrap around the angle of the building. In the morning, when we passed, it smelled of stale beer and sometimes the brewers' draymen could be seen, enormously pot-bellied, purple-faced men, wearing leather aprons, lowering barrels with a rope down a shiny wooden chute from the horse-drawn drays, or else, having completed the delivery, drinking the first pints of the day they were entitled to as 'perks'. (Some of these men drank as many as sixteen pints a day 'regler', according to a Watney's drayman I met in the 1950s.)

And there were shops as minute as the houses – smaller, in fact, in terms of living space because they were houses in which what had been the front parlour had become the shop. They sold things that I was not usually allowed on the grounds that they would be bad for my teeth or my immortal soul, such as what my parents considered to be 'vulgar' comics; more vulgar, but not in the sense of being 'rude', than the *Magnet* and the *Gem*, both of which I was allowed and both of which Kathy enjoyed reading to me as much as I enjoyed listening, much more vulgar than the *Children's Newspaper*, which because of its virtuous nature I already found boring. It was a useless prohibition anyway, as I could always borrow one of these more vulgar comics – 'I say, man, if you let me have a go of your comic you can have a go of my liquorice strip' – from other less watched-over boys at school.

Bad for the teeth were: lemonade made with lemonade crystals, much more delicious in my opinion than real lemonade; toffee sticky enough to pull out entire rows of stoppings; gobstoppers, huge sweets like musket balls that changed colour and the colour of your tongue progressively as you sucked them; sherbet imbibed through liquorice tubes from cylindrical yellow packets that looked like fireworks (oddly enough I was allowed liquorice); toffee apples that always had a thin layer of dust on them that had blown into the shop from the street outside.

Embedded in the pavements at some of the street corners there were cast-iron bollards, shaped like muzzle-loading cannon with imitation cannon balls stuck in their imitation muzzles, against which old men could usually be seen leaning, wearing cloth caps, white silk mufflers or red-spotted necker-chiefs and suits of what even to me seemed antique cut.

In 1927 the poor looked much poorer than they do today, in Hammersmith or in any other part of London. Their everyday clothes in those days, before sponging and pressing and dry-cleaning became commonplace, looked as if they had been slept in. For working men, manual labourers who lived by the sweat of their brows, there was no such thing as winter or summer clothes. A working man wore the same suit all the year round, except on Sundays. In summer, if it was really hot, he might discard his jacket, hardly ever his waistcoat, even though the cloth from which such a suit was made was often thicker and heavier than that used to make a present-day overcoat.

Because of this the poor often smelt. It was not a term of derision as it usually was at Colet Court. 'Yah, you smell!' Although some boys there did smell. It was a fact. One only had to travel, as I sometimes did to my great delight, on one of the tall, two-storeyed tramcars that used to sway down King Street from Hammersmith Broadway with bells clanging, like sailing ships rolling down to Rio, on my way to visit my Auntie May at Stamford Brook, or else travelling down Shepherd's Bush Road en route with Ellen to visit an uncle of hers who had a boot repairing business in Goldhawk Road, to know this smell for yourself, a bitter-sweet odour that a modern traveller, Laurens Van Der Post, identified some thirty-five years later (in connection with the Russian proletariat, en masse) as the smell of soiled clothing, left and forgotten in a laundry basket. A bath was a tub half-filled with water from the copper in which the weekly wash was done. A wash was a lick and a promise in the kitchen sink.

I took more notice of the children of the poor than of the grown-ups, because they were nearer my level in the world in terms of feet and inches, and therefore more often confronted me. I remember the boys more than I remember the girls because the boys tended to move about in gangs. I remember them, not all of course, but many, as being pale and thin, some of them almost transparent, so that looking at their faces with the skin drawn tight over them and their cropped, sometimes shaven heads above, I had the impression of being able to see their skulls through the skin.

But, although some were painfully thin, with bulgy, raw-looking knees protruding below the ungainly-looking shorts they wore, cut down from the discarded trousers of their elder brothers, they were tough, as tough as those of their fathers who had survived the war. Among themselves they fought like ferrets, on the pavements, in the gutters, anywhere, not caring what damage they did to one another or their already tattered clothes, putting in the boot, as it is now called, when they were able, employing methods that at Colet Court were regarded as unfair, except by the gangs of bullies whose techniques surpassed anything the poor could think up at that time in terms of the infliction of physical pain. Their parents fought, too.

Their fights were not the lightning affairs their children were adept in, as if they were torpedo-boats racing in to an enemy anchorage and doing the greatest amount of damage in the least possible time and then getting out again. They were lumbering, major actions between what were more like dreadnoughts, that went on until one or other of the participants was rendered *hors de combat*, or until the police arrived. Such encounters in the streets of Hammersmith were awful to watch, at least to me, because the idea of grown-ups of whatever condition, who were presumed to know better, actually fighting one another, pulling great tufts of one another's hair out sometimes if they were women, was unthinkable.

The children of the poor, boys and girls alike, were resourceful. They had to be. Apart from marbles and iron hoops they had scarcely any bought toys; or if they had they must have kept them indoors as I never saw any. They tied lengths of ragged rope to the crossbars of the street lamps and swung on them, or else used them for skipping. In the autumn the boys played conkers, as we did at Colet Court, bashing away at one another's iron-hard, specially cured horse chestnuts on a string until one or other of them broke. They also played various street and pavement games, such as hopscotch, according to the season of the year, marking out the courts with chalk. If I and my friends used to chalk out the same games on the pavements in Riverview Gardens, residents complained and we got ticked off by the porters, who were numerous. What I envied them were the scooters and little carts, made for them by their fathers or elder brothers, using wood from old packing cases and wheels from discarded roller skates. These scooters made what to me was a wonderfully deafening noise as their owners scooted along the pavements or, more daringly, along the road. I thought them far superior to my own bought scooter from Hamleys which, although possibly a little faster, was depressingly noiseless, being fitted with rubber wheels. In the little carts, which were almost equally noisy, they used to pull their younger brothers and sisters, most of whom should have been in prams if they had had prams, all the way from Hammersmith up Castlenore to Barnes Common and back, a couple of miles each way. How these infants survived such jolting journeys is a mystery.

They were brave, too. In summer, when it was high tide at Hammersmith, some of the boys who could not have been more than nine or ten, used to dive into the then indescribably filthy water from the parapet of the bridge, a good fifteen feet above it, and then, never having been taught to swim properly, dog-paddle to the embankment. They used to do this until a policeman, or a policewoman wearing a helmet like an upturned basin, a huge blue serge skirt and big black boots, used to appear and chase them into Hammersmith, still naked, clutching their ragged clothes, but never catching them.

Besides being tough, resourceful and brave they were also, so far as I was concerned, and anyone who travelled the same route to school as I did, extremely nasty. In the summer of 1928, for the first time I was allowed to go to school without Kathy, providing that I travelled by the back street route, avoiding Hammersmith Broadway, in which my parents were always convinced that I would be run over. Sometimes, but not always, I travelled with another boy who lived nearby us, whose parents bound him to the same conditions.

To us, the perils of this back street route were far more real than any of the risks our parents imagined us running in Hammersmith Broadway, such as being knocked down by a bus or tram. It took us through the heart of territory in which the poor and underprivileged lay in wait while on what appeared to be their more leisurely way to their own schools. Travelling with Kathy, herself a member of the working class, I now realized was like having been provided with some sort of *laisser passer*. In any event, she stood no nonsense from anyone and on the only occasion she did have trouble, when what seemed to me a very large boy, one as tall as she was, crept up behind her and pulled her long hair, she gave him such a resounding slap in the face that he went off howling. Now, using the same route, we encountered the enemy, an enemy waiting to jeer at us, shove us about, smash our straw hats in or pinch our caps, according to the time of year.

If there were only a couple, and they were not too large, we used to stand and fight. We were quite good at fighting, in fact, for much of the time we did little else in the breaks at Colet Court, and quite often we succeeded in sending them away blubbing. Surprisingly, for all their ferocity, they seemed less able than we were to put up with physical pain.

If we did win such a victory, however, our triumph was usually short-lived. No later than the next day we would find ourselves the subject of a major ambush by members of the same tribe. This could be very serious unless we happened to be armed with cricket bats, or even school satchels would do if they were sufficiently packed with books, to bash the boys with. Otherwise the only thing to do was to run for it: capture meant torture, even if it was only of an improvised, not very refined sort, which it more or less had to be in an open street. (At Colet Court, where there were places hidden from view in the playground, to which few masters ever penetrated, torture could mean having drawing pins pushed into the palms of one's hands.) Even so, we sometimes arrived at school, ourselves blubbing, with bloody noses and straw hats stove in and, worst of all, late, in which case we were reported. It is perhaps not surprising that when things got really bad we took to crossing Hammersmith Broadway by the forbidden route.

However, these misfortunes were soon forgotten travelling home to Ther Boiler in the evenings on top of a No. 9 or 73 open bus, bombarding other boys on the tops of other buses with peashooters or squirting water pistols at them; whistling at the girls from St Paul's Girls' School, with whom social intercourse while travelling was discouraged; hiding under the canvas covers, which could be put up over the seats in wet weather, to avoid paying the fare; or better still, waiting for the buses of the Westminster or Premier pirate bus companies, that were fighting a battle for survival against the London General Omnibus Company. Their drivers used to shoot ahead of the sedate red Generals at a tremendous rate and scoop up all the customers, so that, when the Generals arrived, which had to observe a time schedule, the passengers were already half way home. The pirates had no ticket inspectors to speak of, and often their conductors used to let schoolchildren ride free.

But these early back street encounters were as nothing compared with the risks one ran when one was older and was required to wear the ludicrous uniform decreed at St Paul's – black jacket, striped trousers, stiff white collar, black tie. In winter, boys who had attained a certain height wore bowler hats and carried rolled umbrellas.

These last two items, although they conferred a certain, barely tangible, status on those who wore them in the company of their fellow Paulines, had the reverse effect on those wearing them all alone, for example in the Hammersmith Bridge Road.

By this time, aged fifteen or so, it did not matter whether I went to school by way of Hammersmith Broadway or the back street route wearing such an outfit. There were just as many possibilities of being elbowed, tripped or jeered at on the bridge itself (a nasty place for 'an encounter'), or in the Hammersmith Bridge Road, by what I now recognized were no longer schoolboys but semi-grown men. I now walked to school from choice, finding it too much of a bore to queue up for a bus at Ther Boiler in what was always the rush hour when I set out.

Although the penalty for not wearing a hat, whether a bowler, a school cap or a straw hat, was quite severe (a beating usually administered by a prefect), I preferred at this age to be beaten rather than draw attention to myself by wearing any of these sorts of headgear, just as by this time I was prepared to risk punishment for talking with girls I knew from St Paul's Girls' School on their way home by bus, some of whom also removed their hats but for a different reason – because they were good-looking and did not want to look like schoolgirls.

The only good thing about this crazy outfit was the umbrella. Less lethal than a cricket bat (with which if you hit someone really hard you might easily kill him), the umbrella, used as one would use a rifle with a bayonet or as an outsize truncheon, rather like those carried by the mounted police, was an ideal weapon.

Uncivilized as my behaviour may seem today, in a more squeamish but much more dangerous age, I can only plead that I had no choice. The penalty of defeat or, even worse, capture, would have been by this time much more serious than anything I experienced at Colet Court.

As a result of this misuse, my umbrella and those of my schoolfellows who found themselves in similar situations soon became useless for the purpose for which they were intended, either failing to open when it rained and they were needed, or else opening and falling to pieces.

These journeys from Three Ther Mansions over the bridge and through the streets of Hammersmith altogether continued for eight years of my life (not including the period when, as a small child, I attended the Froebel kindergarten in Baron's Court). In me they engendered some of the feelings of excitement, danger and despair that some nineteenth-century travellers experienced in darkest, cannibal Africa and in the twentieth century in the Central Highlands of New Guinea. And even today and now with even more reason, I sometimes experience a chilly sensation when walking alone down a narrow south London street.

8

Lands and Peoples

Up to now the reader may feel, and with some justification, that my travels have been of a somewhat parochial nature. By the time I was eight years old, apart from a visit to the Channel Islands in 1923, I had never been out of England, not even to visit Scotland, Ireland or Wales. Yet, in spite of this, I already knew a good deal about these places and their inhabitants, as well as the wider world beyond the British Isles.

This was because, some time in the 1920s, my parents took out on my behalf a subscription to *The Children's Colour Book of Lands and Peoples,* a glossy magazine edited by Arthur Mee, at that time a well-known writer who was also responsible for the to me rather boring *Children's Newspaper,* which I had given up buying in favour of more trashy, exciting comics. *Lands and Peoples,* according to the advertisements which heralded its publication, was to come out at regular intervals and when complete the publishers would bind it up for you to form six massive volumes.

What distinguished *Lands and Peoples* from other similar publishing ventures was that it was illustrated with colour photographs. 'Marvellous as photography is, in one sense it has failed,' the editor wrote. 'The colour photograph is a dream . . . Here it is that the bold idea of *Lands and Peoples* has succeeded beyond all expectations. What Science has failed to do, Art has done . . . Some seven hundred photographs from all over the world have been coloured by artists from original sources, so that they become actual photographs with colour true to life, a remarkable anticipation of the final triumph of the camera that will show the world as it really is, in all its glow of red and green and gold.' It was no hollow claim. The colours were in fact more true to life than those produced from much of the colour film in use today.

What excitement I felt when *Lands and Peoples* came thudding through the letter-box at Three Ther Mansions, in its pristine magazine form. To me, turning its pages in SW13 was rather like being taken to the top of a mountain from which the world could be seen spread out below, a much more interesting world than the one my father read bits out about from his *Morning Post* to my mother while she was still in bed, a captive audience, sipping her early morning tea.

Lands and Peoples was addressed to 'The Generation whose business it is to save Mankind' – mine presumably, although I had no suggestions to make about how it should be done. 'We are marching,' Mee wrote, an incurable optimist, apparently envisaging a pacific version of the Children's Crusade, with the children waving white flags instead of brandishing crucifixes, 'towards a friendlier and better world, a world of love in place of hate, of peace instead of war, and one thing is needful – an understanding of each other . . . it is the purpose of this book so to familiarize us with the lands and peoples of the world that we can cherish no ill-will for them. We are one great human family, and this is the book of our brothers and sisters.'

I never read the text. All the efforts of what was presumably an army of experts in their various fields, painfully adapting their ways of writing to render them intelligible to infant minds, were totally wasted on me. All I did, and still do, for I still have the six volumes, was to look at the photographs, of which there were thousands in black and white besides the seven hundred or so in colour, and read the captions and the short descriptions below them of a world which, if it ever existed in the form in which it was here portrayed, is now no more. What I saw was a world at peace, one from which violence, even well-intentioned sorts, such as ritual murder and cannibalism, had been banished, or at least as long as the parts kept coming.

In Merrie England, donkeys carried the Royal Mail through the cobbled streets of Clovelly, milkmaids in floppy hats churned butter by hand in the Isle of Wight, swan uppers upped swans on the River Thames, genial bearded fishermen in sou'westers mended nets on the Suffolk coast, town criers rang their bells, and choirboys wearing mortarboards, using long switches, beat the bounds of St Clement Danes.

In the Land of the Cymry, which is how *Lands and Peoples* described Wales, ancient dames wearing chimney-pot hats stopped outside their whitewashed cottages to pass the time of day with more modern, marcel-waved neighbours, and cloth-capped salmon fishers paddled their coracles on the River Dee.

In Bonnie Scotland, brawny, kilted athletes tossed the caber, border shepherds carried weakling lambs to shelter from the winter snows, women ground corn with stone hand-mills in the Inner Hebrides and, far out in the Atlantic, on St Kilda, the loneliest of the inhabited British Isles, men wearing tam-o'shanters and bushy beards were photographed returning homewards with seabirds taken on its fearsome cliffs.

In the Emerald Isle, bare-footed, flannel-petticoated colleens drew water from brawling streams, or knitted socks in cabins in Connemara, while little boys in the Aran Islands wore skirts to protect them from being kidnapped by the fairies.

Further afield, as I turned the pages, covering them with Bovril (for one of my greatest pleasures was to go to bed early with *Lands and Peoples* and at the same time eat Bovril sandwiches), I came upon Czechs and Slovaks and Hungarians wearing embroidered petticoats irrespective of sex, and incredible hats – in Rumania there were men who wore garters with little bells on them, who looked like sissies to me. I also saw fellers of proud giants in the Canadian forest; savages of New Guinea – their hair plastered with grease and mud; Jews in Poland – a new state with a glorious past; sun-loving Negroes in South Africa; happy Negro children romping in the 'Coloured Section' of New York; Kirghiz tribesmen on the 'Roof of the World'; Orthodox scholars wearing arm thongs and phylacteries studying the ancient laws of their people; penguin rookeries in the Great White South; geishas negotiating stepping stones in Cherry Blossom Land; hardy Indian women on beds of nails at Benares; laughter-loving girls of the Abruzzi; Flemings and Walloons – little Belgium's two sturdy races; Macedonian women weaving fine cloth with deft fingers; haughty-looking redskins decked in eagle's feathers and wampum; Germany – rich country of an industrious nation; and so on, and so on.

Everywhere in *Lands and Peoples* there were photographs of schools and schoolchildren; ruinous-looking schools in which little Negroes were learning to add up, schools in Japan with the pupils marching round in circles dressed in military-looking uniforms, clinical-looking schools in the United States in which the children were being inspected for dental decay, children brandishing enormously long slates in Burma, memorizing the Koran in oases, going to school in wheelbarrows in China, learning to read in lonely Labrador and being bastinadoed in Persia, one of the only examples of violence in the entire work. That children had to put up with going to school in other countries besides my own I found encouraging. And here at home they didn't have the bastinado, even at Colet Court.

Each time I got through all six volumes of the *Children's Colour Book of Lands and Peoples* I began again, in much the same way as the Scotsmen of Bonnie Scotland, sturdy independent folk, who painted the Forth Bridge, would start all over again as soon as they had applied the final brush strokes to that miracle of Scottish engineering.

The other work of travel, which I also smeared with Bovril and which for many years remained my favourite book to the exclusion of almost all others, was *Travel and Adventure in Many Lands* by Cecil Gosling, formerly His Majesty's Envoy Extraordinary and Minister Plenipotentiary to the Republic of Bolivia. I received his book, which Methuen published in 1926, as a present from my father on my ninth birthday in 1928. It was heady stuff. Gosling had a wonderfully exciting career, principally in South America, where he had travelled extensively, usually on horseback. In the course of these journeys he had encounters with robbers, hordes of wild pigs with razor-sharp tusks, vampire bats, poisonous insects, including horrible, hairy spiders and, for me most terrifying of all, with the Paraguayan pirana fish, his description of which at one time I could quote word for word, and which if the demand was sufficiently insistent, I used to recite to my friends at school during break, one of my major social accomplishments. 'Perhaps the most dangerous inhabitant of Paraguayan waters,' he wrote, 'is the pirana, which I suspect of causing more loss of life to man and beast than alligators, *surubis*, or *mangarujus*, and such aquatic monsters,' whatever they were. 'The pirana is a scaled fish, similar in shape to our European perch, only the

77

head is of more aggressive appearance, the mouth being armed with a most formidable set of teeth, more like those of a wild animal than a fish. With these, and aided by his powerful jaws, there is literally nothing that he will not bite through. I have put a lead pencil into the mouth of one, and seen it bitten clean off, and I have also seen him bite off the edge of a keenly tempered knife. His ferocity is equally developed, and one has to be very careful about handling these demons when caught, and when getting the hooks out of their mouths. When taken out of the water they make a barking noise . . .

'A friend of mine, a police inspector, while bathing, was attacked by a shoal of them, which mutilated him in such a manner that he at once swam back to the bank to the spot where his clothes were, picked up his revolver and blew out his brains.'

Reading this passage or reciting it to an audience always used to make me feel cold inside, although I had not the slightest idea at that time what the particular mutilations might be that had caused the police inspector to blow his brains out.

But it was not only the blood-and-thunder that recommended His Majesty's Envoy Extraordinary and Minister Pleni-potentiary to me. There were other aspects of his life which were almost equally fascinating, not to say unbelievable, as, for example, the strange encounter he had while travelling in a Pullman car from Mendoza, a town in Bolivia, over the Andes to La Paz in Chile, sometime before 1914. Seated at the far end of the car was a gentleman of about fifty years of age who kept on staring at the author in a way which he was beginning to think rather ill-mannered, when all of a sudden the stranger got up and came towards him, raising his hat which he presumably kept on his head for the duration of the journey, and addressed him in French.

'"Pray, excuse my addressing you," he said, "but I feel sure that you are some relation of my old friend Gyldenstolpe!" It was Count Prozor, the Russian Minister to Brazil, who must have had a peculiar facility for seeing likenesses, for his "old friend" was my uncle who had been dead for about twenty years.'

It was such anecdotes, recording the pleasures as well as the horrors of travel, that drew me to Cecil Gosling and that make me, to this day, one of his most loyal and enduring readers.

9

Mystery Tour
(1927)

In the summer of 1927, I went on holiday with my parents to Swanage, a smallish seaside resort on the Dorset coast, between Bournemouth and Weymouth. We travelled by train from Waterloo, and as soon as my father had made sure of our reserved seats in the compartment, we took our seats in the dining-car. I had already been allowed on the footplate of the engine, hauled up on to it by the driver or the fireman, both of them grimy, friendly men. In those days it was commonplace for small boys to visit the crew and be shown the engine before setting out on a journey.

I remember the snowy-white, ferociously starched tablecloths that began to be spotted with specks of black from the engine as soon as the train started, the smell of food, mixed with the acrid smell of train, itself a compound of oil and burning coal and other unidentifiable ingredients, and the musical tinkling of glasses and cutlery as we picked our way among the points, past a platform marked Brookwood, from which my father told me special funeral trains set out for what was England's largest cemetery, some thirty miles away in the country.

'Why,' I asked, 'did they take dead people all that way when they could be buried in a cemetery where their houses were? Didn't they have cemeteries?' We had cemeteries near us in Barnes, as I well knew.

'Brookwood,' my father replied, 'is for well-off people,' which, although he seemed to think this a sufficient answer to my question, did not really answer it.

The train roared on through deepest Surrey (I found later on, when I was given a bicycle and explored it, that a lot of Surrey was not really country at all, being covered with firing ranges, barracks, golf courses and lunatic asylums), while the soup served by one of the resplendently uniformed dining-car attendants slopped over on to the recently spotless tablecloth however much I tried to stop it.

'That's where the crematorium is,' my father said as we flashed past a small railway station and a lot of pines and tombstones.

'What's a crematorium?' I asked.

'I'll tell you when you've finished your lunch,' he replied.

It was on this train, too, that I first experienced the pleasure of looking down a railway lavatory pan and seeing the permanent way rushing past beneath me, something that has never ceased to fascinate me, whether on the Orient Express, the Trans-Siberian Railway, the 12.15 to Ernakulam Junction, or the 16.30 to Penzance.

Every day that summer at Swanage, if the weather was fine, morning or afternoon according to the state of the tide, we went to the beach where my father had hired a bathing tent for the duration of the holiday. The tent was situated at what was widely regarded as 'the better end' of the beach, beyond the Beach Café, inland of which some of the more hideous urban developments were taking place. The beach near the centre of the town, where stood a monument to a naval victory by King Alfred over the Danes in AD 877 (crowned, rather surprisingly, with iron cannon balls) was equally widely regarded as being the place where the 'day trippers', who used to come over from Bournemouth on the paddle-steamer, congregated. It was at the Beach Café the following year that, in my father's absence on a journey with the coat and suit collection in the North, my mother and I became hooked on what were then in Britain the new-fangled Bell Fruit Machines with such disastrous results that we voluntarily signed a document in which we promised to renounce them for ever. Much earlier my mother had signed a similar document in which she promised not to order anything from Harrods other than household necessities without first consulting my father.

It was while on the way to the 'better end' from what was presumably the 'worse end' that we used to pass the place where motor charabancs were parked and their proprietors used to lie in wait for customers. They advertized their various excursions on blackboards, using coloured chalks, sometimes, if they had the facility, making little vignettes of the principal attractions – 'Ancient Corfe Castle', 'Lovely Lulworth', 'Weymouth, Naples of the West', 'Historic Wimborne – See the Famous Chained Library in the Minster', and so on. And always, without the vignette but some times with the insertion of a large question mark, 'A Mystery Tour', with an opportunity for partaking of a 'Real Dorset Cream Tea' (not included in the price of the ticket).

Eventually, by going on long enough about it, I managed to persuade my parents to take tickets for the mystery tour. To me their comparative lack of enthusiasm for this project, the thought of which threw me into transports of excitement, was a slight surprise, especially as both of them loved the open air and what my father called 'a good blow' in an open motor car. They would certainly get a sufficiency of air in a charabanc.

It was not until much later that I realized that neither my father nor my mother, while inhaling their quota of fresh air, wished to do so in the company of large numbers of other human beings. Just as I did not realize at the time that both of them were bored to death by beaches and that the hired tent and our daily visits to what sometimes amounted to only a few feet of sand between it and high water (none at all if it was a spring tide), was solely for my benefit. And it was my mother who told me this years later, goaded into doing so by some unkindness on my part, and at the same time telling me that my father had disliked every seaside holiday he ever had and was only really happy when he was sculling on the Thames.

We took our seats in the charabanc. I was practically delirious with joy. A charabanc was originally a horse-drawn vehicle of French antecedents, with banks of seats one behind the other, hence its name. A motor charabanc, apart from its seating arrangements bore, unlike an electric brougham, comparatively little semblance to its horse-drawn predecessor. Motor chara-bancs looked like enormously inflated versions of open touring motor cars, painted in more powerful colours – this one was bright red. They were fitted with folding canvas hoods which could be put up if the weather turned nasty but with the same sort of difficulty, if there was a strong wind, that when I became a sailor I associated with the handling of a royal or an upper topgallant in similar circumstances. The banks of seats in motor charabancs became progressively higher towards the back, as in some theatres, so that the passengers there, in the equivalent of the pit, as it were, would not miss anything. It was the rearmost seat that I 'bagged' for the three of us.

Watching the charabanc fill up I began dimly to apprehend, in spite of my excitement, some of the reasons that might have contributed to my parents' apparent lack of keenness for this outing, although they were now putting a brave face on it. The other passengers, at least the majority of them, were trippers, whether day trippers or other sorts of trippers it was impossible for me to say; but trippers they undoubtedly were. They were certainly not the sort of people who had their own bathing tents hired by the week, month or fortnight. In fact, a number of them looked exactly like the characters depicted in the 'rude' postcards on sale near the sea-front, of which my parents used to express abhorrence, rushing me past them, but which, nevertheless, I discovered, they used to send to their more frivolous acquaintances 'in the better end of the business' under plain cover, or else used to keep as souvenirs hidden in drawers at Three Ther Mansions, where I subsequently discovered them, together with even more interesting material.

In spite of their appearance the trippers were both jolly and friendly, much more so than some of our neighbours at the 'better end'. They asked me if I was enjoying myself and being a good boy, called my father 'governor', my mother 'mum' and the charabanc (which I called a 'sharrabang', believing the name to be something to do with the noise it made), a 'sharrer'.

Then we were off. By now I had found myself a more desirable seat, immediately behind the driver who was large and easy-going and who allowed me to chatter away to him – there were no prohibitions about talking to the driver in those days. As soon as we got out of the town he allowed me to honk away on the rubber bulb of the motor horn, telling me when to do it when we rounded the bends.

At that time the official speed limit for such vehicles as charabancs was still as it was for cars. God knows what speed we actually attained in our red 'sharrer' that warm August afternoon; but whatever speed it attained it was sufficient to make some of the more timid female passengers utter shrieks of fear as we roared round the bends, horn blasting, and to make other passengers complain about the head-on collisions they were having with various sorts of insects which, as well as making a nasty mess of their clothes, were quite painful if they were hit in the face by them at such a velocity. This did not matter to the driver, whose name was Fred, to me, or to the other passengers in the front row of seats, since we were protected by a huge expanse of what in the event of an accident was still, in 1927, lethal plate glass.

At Corfe Castle Fred came to a halt sufficiently long enough to address us through an old tin megaphone.

'This is Corfe Castle,' he said, 'and that is the castle,' as if we could miss seeing the vast looming ruin and making me wonder if he was 'all there'. 'We shall be stopping here for forty minutes on the homeward run. This allows ample time to inspect the ruins and partake of a Dorset cream tea at the Castle Tea Rooms,' which seemed pretty silly, telling us all this, when we were supposed to be on a mystery tour. Anyway *we* had already visited Corfe Castle under our own steam, walking the four and a half miles or so to it from Swanage by way of Nine Barrow Down, travelling back by train afterwards. *We* had no need to visit the ruins with the trippers. *We* could concentrate on the cream tea.

From Corfe Castle our tour took us westwards along the foot of the Purbeck Hills. Eventually we crossed them by a narrow serpentine road, the charabanc leaving behind it a long plume of white dust, and entered an entrancing region where the wild Purbeck heathland mingled with more gentle, domesticated country in the long, narrow enclave of the valley of the River Frome, a singularly beautiful stream which has its source high in the North Dorset Downs. To the north of this valley was yet more heathland, Hardy's Egdon.

To the left of the lane, and to the right of it too, was heath, but already partially tamed, with little fields, some with ponies in them. There were also spinnies of young leaf-bearing trees among the pines and the blazing yellow gorse, like parts of the New Forest, of which I had enjoyed almost equally fleeting glimpses travelling from Waterloo to Swanage on the train.

Just before Wareham Fred turned off into a long stony lane, which he said was called Holme Lane. He was still going at such a rate that when we came to a hump-backed bridge over the branch line from Wareham to Swanage, he hit the hump such a crack that it was a miracle we were not all thrown out, and this elicited a few more shrieks from the ladies, although no one actually complained. Probably no one on board, including my parents who, although they had both driven, were certainly not competent to do so, knew much about how to drive a motor car, let alone a motor charabanc. Not that I cared, sitting behind Fred, honking the horn when he told me to, watching what I thought was the most beautiful lane I had ever seen unroll in front of me, between plantations of silver birch and pine and through tunnels of what had been until recently a riot of mauve rhododendrons, past green fields with cattle grazing in them, and a park with huge old trees, oaks and beeches and cedars growing in it, with a gentleman's residence in the background, before which Fred momentarily halted his charabanc to give us the opportunity to admire it. Beyond it, to the right of the road (although I did not see them until years later while on a cycling holiday), were the lush water meadows of the Frome, with water-mills and fish-ponds and weirs and sluices and the site of a medieval church and priory, and here and there the sort of cottage I had always imagined King Alfred burning the cakes in, ever since receiving the illustrated *Nursery History of England* as a Christmas present. And to the left were the Purbeck Hills, fleetingly glimpsed across dark expanses of heathland.

Eventually, having driven us through a couple of water splashes, throwing up clouds of spray and churning up the gravel in the process, and thereby giving cause for more shrieking, Fred deposited us on the threshold of the first mystery. He was a terror, Fred. He drove like a mad Persian I met years later in Meshed. The place where this mystery was to be unveiled seemed highly suitable, a point where the lane, close to the River Frome now on the edge of its water meadows, skirted a spooky-looking wood. Here was a gate with a drive beyond and for a moment when he stopped, before the passengers found their voices and Fred found his megaphone, there was a silence broken by the hum of insects and the semi-somnolent sound of wood pigeons.

At the end of the drive there was what I would now identify as a Gothic gatehouse, with a pointed entrance arch, window tracery and mock battlements, but at that time described to me as being like the wood, spooky, which was what the builder no doubt intended.

Any suggestion of spookiness, however, was soon dispelled by Fred with his megaphone, which sounded more like a foghorn in these otherwise peaceful surroundings.

'Ladies and gentlemen,' he said, 'this is a Beauty Spot. The ruins of Bindon Abbey. The Abbey was built, 1172, by monks and knocked down at the Reformation. Among the ruins you will see the open grave in which Angel Clare laid Tess, having carried her here from Wool Manor House on the other side of the Frome, as written about in his famous novel *Tess of the d'Urbervilles* by Mr Thomas Hardy, now living at Dorchester, whose house you can see if you take our Hardy Country Tour. As this is a Beauty Spot we will stop here for fifteen minutes to allow you ample time to view the ruins and the outside of the house which is not open to the public. Likewise, the picking of flowers, wild or otherwise, is prohibited. Bindon Abbey is the property of Mr Weld of Lulworth, a member of the well-known Roman Catholic family. Admission tickets, price twopence, may be obtained at the gatehouse. I will sound my horn, once only, when it is time to leave.'

What can I say after all these years of my first visit, and the last for what proved to be many years to come, to the ruins of Bindon Abbey, a Cistercian monastery, destroyed after the Reformation in 1539, coming eventually to the Welds of Lulworth by purchase from an Earl of Suffolk, to whom it descended from Lord Poynings, who received it at the Dissolution? When I saw it as a still-small schoolboy it was as it had been for centuries already, the remnant of a ruin, most of which had been carted away to build other buildings, of which little more than the foundations remained: no cloister, no nave, no refectory, no dormitory, a few bases of columns, a few tombs, everything eaten up by ivy, or strangled and convulsed by the roots of trees, hidden from the outside world by woods.

What do I remember most? The open sarcophagus in which Clare, one of the arch prigs of English literature, laid Tess; the bases of the columns, the ivy-clad outlines of the abbey, vaguely delineated? Much more than any of these I remember the vistas of the long ponds in which the monks may have bred their carp, the sound of water running through the sluices, the rustle of the leaves and the melancholy, even angry, cawing of the rooks, galvanized into activity by Fred with his appalling megaphone.

But what I remember more than anything was the feeling of the actual, physical presence of the monks themselves, who had been dead for more than three hundred years, as I walked the paths alongside the dark waters, until the sound of Fred's motor horn and my mother's cries of 'Eric, where are you?' brought me running back.

10

Something in West One
(1936–8)

At St Paul's I became a Scout in order to avoid being drafted into the Officers' Training Corps, the OTC, which would have meant wearing an insufferably itchy uniform that was blanket thick. As a Scout I learned to light a fire with one match and to use a felling axe without dismembering myself. On Saturdays we engaged in bloody night battles with other troops of Scouts in the swamps of Wimbledon Common. In the holidays we went camping in the beautiful parks of gentlemen's country seats. Because I got on with Jews – I still do, I think they find my lack of subtlety restful – I was given command of a Jewish patrol. One of them, who was as prickly as a present-day Israeli, refused to wear Scout uniform and appeared at our open-air meetings wearing a double-breasted overcoat, a bowler hat and carrying a rolled umbrella. Another, who used to charge his mother half a crown to kiss him good night, was always loaded down with silver. On one occasion, when there was a danger of our losing one of the outdoor games known as 'wide games' (which necessitated covering large tracts of ground on foot) another member of my patrol whose family owned a huge limousine in which he used to arrive at wherever was the meeting place, summoned the chauffeur who was parked round the corner, and six scouts whirred away in it to certain victory.

In the spring of 1936, when I was sixteen years old, my father announced that, as it appeared unlikely I would pass the School Certificate Examination (the then equivalent of O-levels), in mathematics, a subject in which it was obligatory to pass, he had decided to take me away from St Paul's at the end of the summer term and 'put me into business'. I did fail. I was sorry about this decision. I was good at English, History, even Divinity, and I had dreamed of reading History at Oxford.

Apart from an innate inability to cope with mathematics the

only disadvantage I laboured under at St Paul's, and being a Scout made not the slightest difference, was that I had a curious sense of humour which meant that if anything came up in class with a suggestion of *double entendre* it caused me to dissolve into hysterics, for which I was punished, sometimes quite severely. In other words I had a dirty mind.

For instance, on one occasion, when we were reading Scott's *Marmion* aloud, it became obvious to myself and everyone else in the classroom that by the working of some hideously unfair natural process of selection it would fall to me to read a completely unreadable part of the romance in Canto Two, entitled 'The Convent', which concerned the blind Bishop of Lindisfarne. And you could have heard a pin drop when I got to my feet.

> 'No hand was moved, no word was said
> Till thus the Abbot's doom was given.
> Raising his sightless balls to heaven . . .'

was all I could manage before going off into peals of mad laughter and to be beaten by John Bell, the High Master, a hedonist who showed in as marked a way as possible in the circumstances where his sympathies lay by beating me as hard as anyone else sent to him for punishment, and then giving me a shilling. I have never forgiven Scott.

My childhood was at an end. I had no complaints. I had been an only child and what was certainly an excess of loving care and attention had always been lavished on me. I had found it difficult to excuse my parents' visits to me in their motor car with Lewington at the wheel when I was away camping, something my fellows did not let me forget. It is true that I suffered from a sense of inferiority, an inability to be 'good at figures', as my father put it, and a marked reluctance even to attempt to row seriously and be chosen for the school eight and row at Henley, for which I had both the style and the physique, having learned to row before I learned to scoot a scooter. These were the only fields in which my father really wanted me to excel, fields in which I, too, would like to have excelled but somehow felt incapable of doing so. Thinking about them I felt like the man with the inferiority complex who is told by his consultant that he really is inferior. There was no one else I could blame. I was only sorry that I had disappointed my father. My mother didn't care whether I excelled in them or not.

During eighteen months spent learning business methods, I survived several office purges of the kind that take place frequently in advertising agencies when they lose an important account. I survived them not because I was astute but simply because I was paid so little as 'a learner' that there was not much point in giving me the sack. However, when the agency lost an important breakfast cereal account, and when all my best friends went overboard with it, I decided that I too wanted to go overboard with them. I had had enough of learning business methods.

That August in 1938, with an international crisis building up in Europe over Czechoslovakia and the Germans mobilizing, I went on holiday to Salcombe in south Devon. Diving in Starehole Bay near Bolt Head, I saw what remained of the four-masted Finnish sailing barque *Herzogin Cecilie* of Mariehamn, which had crashed into the Ham Stone in April 1936 with a cargo of grain from Australia on board, eventually becoming a total loss. And on the way back to London, while changing trains at Newton Abbot, I wrote a letter to the owner, Gustav Erikson, in Mariehamn, asking him for a place in one of his grain ships.

To become an Erikson apprentice I had to be at least sixteen years of age (birth certificate required), of strong constitution (two doctors' certificates), and of good moral character (one clergyman's certificate, which he signed without setting eyes on me). If I died from natural causes, by falling from the rigging, or by being washed overboard, my father would get back a proportion of the £50 ($250) which he paid to make me an apprentice. I was also to be subject to Finnish law and custom. My wages were to be 150 Finmarks a month, at that time the equivalent of 50p ($2.50). Even the captain only received about 4000 Finmarks (£20 or $100).

On 16 September, at the height of the Munich crisis, the day after Chamberlain visited Hitler at Berchtesgaden and a week after the French had called up their reservists, by which time it was beginning to look as if it was not going to be a particularly good year to be at sea in an engineless sailing ship, my father received the following letter:

H. Clarkson and Company Limited

52 Bishopsgate,
London EC2.
15 September, 1938

George A. Newby, Esq.,
Messrs Lane and Newby,
Wholesale Costumiers
   and Mantle Manufacturers,
54, Great Marlborough Street,
London W1.

S/V *Moshulu.*

Dear Sir,

We now have a letter from Captain Gustav Erikson advising us that he wishes your son to join this vessel which is discharging at Belfast on the 26th September.

If you will now send us the £50 [$250] premium, we shall send you a contract for his service in this vessel . . .

As your boy will be arriving in Belfast in the early morning on the steamer from Heysham he will be able to go direct to the ship which is discharging her cargo in York Dock.

Yours truly,
For H. Clarkson and Co. Ltd.
A. S. Calder

I Go to Sea
(1938)

The following is a letter from Mr W. H. Eyre, otherwise known to his associates as 'Piggie'. Solicitor, notable oarsman (he rowed in the Thames Rowing Club crew that won the Grand Challenge Cup at Henley in 1876), Mr Eyre was an old friend of my father.

Rye and Eyre,
Solicitors,
Golden Square,
London, W1.   12 September, 1938

Dear George
        . . . it surprises me that Eric wants to be a sailor, and I do strongly hope that he will reconsider it, as I am certain his mother is *sure to miss him dreadfully*, as you will also. You say he is to be apprenticed, but I think he would give it up unless the conditions are very far better than I can remember when I was a small boy of the voyage out to New Zealand, and two years later when we returned in a seven-hundred-ton barque, a most stormy voyage which lasted nearly six months! We were nearly wrecked coming round Cape Horn (southernmost tip of South America), and had to put in at the Island of St Helena for provisions. I was nearly ten years old then, it was 1858, and my brother and I used to go up the masts and out on the yards and do a lot of little odd jobs for the skipper and the mates. We knew every rope in the ship and lots of 'shanties'. The apprentices' life was a *very hard one* and the poor fellows used to get a rope's ending from our severe old skipper (a very painful operation), but I suppose there will be nothing like that on the 'line' you mention. But he will be away

continually, and subject to continual risks so, as he is your only child, I do hope that you will arrange for him to have a trial voyage before he becomes bound to serve five years, or whatever the term is. Talk it over with Mrs Newby and then, both of you, do all you can with the boy. Surely if he does not like commerce there are many occupations in which he would probably do well. You might even article him in some good solicitors' firm where he would get plenty of time for rowing and all other amusements whilst serving his time – or a chartered accountant is a good 'trade' to learn, as they are fit for all kinds of well-paid Government and other posts. As are also properly trained engineers, either civil or mechanical – but you probably know as much or far more than I do about starting a clever young fellow in life.

Forgive my bothering you with all this but I cannot help feeling very strongly that, for your's and his mother's sake, he should be kept in England.

The next letter is from Mimi, ex-fashion buyer, the sister of Beryl and Mercia Bamford, and one of my mother's greatest friends:

Barton-on-Sea, Hants
Friday, September 1938

Dear Pom [an abbreviation of my mother's maiden name, Pomeroy]

I have just been talking to Beryl and Mercia on the telephone and Beryl told me Eric is going to sea. Pom, dear, I can imagine how you feel, but you have always been a sacrificing soul, and this may seem like the *supreme sacrifice*; but you can and will do it you know, Pom, and it thrills me to think of the life he will lead and what he will see of this world, and I can sympathize with Eric. He was always an 'adventurous spirit', always full of zest and a great little reader, with such imagination. Do you remember that odd-shaped stone he found when he was quite small, with holes in it at Branscombe, which he said looked like a 'prehistoric aunt'. If the opportunity comes his way he could even be a great traveller and writer. Give him my love and tell him he must take some good reading with him. I enclose a little something for this.

What times we live in, but what fun we've had together!

We've just had our gas masks fitted. Should there be air raids with Southampton and Calshot round the corner we're not in the safest place; but where is safe?

Don't take the parting too badly, old girl. He has his life to live, and what a father and mother he has had, and what a childhood!

Yours ever,
Mimi

Here is my first letter on board ship to my parents: eighteen years later this and subsequent letters to them provided me with the necessary material for the opening chapters of *The Last Grain Race*:

S/V Moshulu, East Side,
York Dock,
Belfast

26 September 1938

Dear Mummy and Daddy,

... I was up on deck on the steamer from Heysham about 6.30 just in time to see the terrifically high masts of the *Moshulu* rising high above the dock sheds and looking very cold and remote in the early morning. After breakfast in the steamer I took a very ancient taxi that was practically falling to pieces to the ship and when it arrived alongside it was so big that I felt like a midget. It is more than three thousand tons and is the biggest sailing ship in commission in the world.

I went up a gang plank and spoke to a very tough-looking boy with slant eyes like a Mongolian who was oiling a donkey engine, and asked him if he would help me with my enormous trunk [made by Louis Vuitton and bought in a second-hand shop, for £4 or $20]. He picked it up, having made threatening gestures at the taxi man who was trying to overcharge me, and carried it up the plank on his back all by himself!

Meanwhile, sacks of grain were being hoisted out of the hold and weighed before being taken ashore and into the sheds – altogether the ship brought back more than sixty-two thousand sacks from South Australia on this last voyage and was a hundred and twenty days at sea, which is rather slow. A hundred days from Australia to Britain is good, anything under a hundred, very good.

Then my new friend, Jansson, who comes from the Åland Islands, took me into the starboard fo'c'sle where I am to live until the watches are appointed, which will be on the day we go to sea. It is about thirty feet long and about thirteen feet wide, with wooden bunks from floor to ceiling, one above the other, like coffins with open sides. I am not sure how many but will tell you later. The boys seem pleasant enough, but not exactly gushing. About three-quarters of them speak only about a dozen words of English and some of those are swear words, which is twelve words more than I speak of Swedish which is the language in which all orders are given, rather than Finnish, which would be too difficult for non-Finns, I suppose. Swedish/ Finns from the Åland Islands and Finns make up the majority of the crew.

They gave me some coffee and then I was told the second mate wanted to see me on the bridge deck, which is amidships above the fo'c'sles, where the ship is steered from, not from the poop. He was a pale, thin fellow and after asking me my name suddenly said, 'Op the rigging!'

I simply couldn't believe this. I thought they would give one a bit of time to get used to being in a ship. I was wearing my Harris tweed jacket, grey flannel trousers and those leather shoes with slippery soles which I took the nails out of because I thought they might damage the decks! He wouldn't allow me to change, not even my shoes, just take off my jacket and shirt. I was in a sort of daze. I swung out over the ship's side and started to climb the ratlines, wooden rungs lashed to the shrouds which hold the mast up, quite wide at the bottom but only a few inches wide when you get under what is known as the 'top', where it was difficult to get feet as large as mine on to them. The top is a platform and to get on to it I had to climb outwards on rope ratlines, like a fly on a ceiling, and when I got on to it, looking down I almost fainted.

I thought this was enough, but 'Go on op,' he shouted, and so I went on 'op' by rope ratlines, some of them very rotten after the long voyage – you have to hold on to the shrouds, not the ratlines, in case they break. These ratlines ended at the cross trees, which are made of steel and form a sort of open platform, like glass with my slippery shoes on, a hundred and thirty feet up with Belfast spread out below.

'Op to the Royal Yard,' was the next command. This was forty feet or so of nearly vertical and very trembly ratlines to just below the Royal Yard, the topmost one, on which the Royal Sail is bent at sea. Here, there were no more ratlines, and I had to haul myself up on to it, all covered with grease from something like a vertical railway line [a mast track on which the yard is raised and lowered by a halliard] on the face of the mast.

This yard was about fifty feet long, and made of steel, like all the other yards and masts. It had an iron rail along the top (a jackstay) to which the sail is bent and underneath it is a steel footrope [the 'horse'] on which my shoes skidded in opposite directions, so that I looked like a ballet dancer doing the splits. At the yardarm I was – I learned the heights when I got down – about a hundred and sixty feet over the dock sheds, which had glass roofs – what a crash, I thought, if I fell off! It was better looking further afield. There were marvellous views of the Antrim Hills and down Belfast Lough towards the Irish Sea.

Back at the mast, thinking thank goodness that's over, I heard the mate's voice telling me to climb to the very top of the main mast [the main truck]; but I couldn't do this with such slippery shoes on, as there were only two or three very rotten-looking ratlines on the stays [seized across the royal backstays], and then nothing but about six feet of absolutely bare pole to the cap. So I took off my shoes and socks, which were even more slippery than the shoes, left them on the yard and then shinned up, past caring what happened, more frightened of the mate than anything else, until I could touch the cap which was like a round, wooden bun. This cap I have discovered is nearly two hundred feet above the keel, much higher than Nelson's column [which is only a hundred and forty-five feet high].

Now he was shouting to me to sit on the top; but I would have rather died than do that and so I pretended not to hear him. When I got down he was angry because he said it was unseamanlike to take off my shoes and socks, and also that the shoes might have fallen and injured somebody. I think the part about sitting on the cap must have been a joke, because he never mentioned it and none of the others have done it. Then he sent me to clean the lavatories; but I was allowed to change first. My trousers got awfully dirty with Belfast soot up in the rigging and will have to be cleaned.

I am not writing this to worry you but only because I feel that I will be alright in the rigging from now on. It was probably a good idea to get it over. Anyway, it is much better than cleaning the lavatories.

Please do not send anything but letters and money as I have no room for anything in my bunk except myself.

This evening we are going ashore for what the boys call a 'liddle trink' at a pub called the Rotterdam Bar, a farewell party for the boys who are leaving the ship and going home. I never imagined that one day I would go to a pub called the Rotterdam Bar and be a sailor.

All my love,
Eric

The following day, 27 September, the Royal Navy was mobilized. However, cleaning what had become 'my lavatories', taking further steps in the rigging and over the side chipping rust from the hull most of the day, I had little time for reflection about what was going on in the world outside. And as soon as I stopped work I immediately fell asleep.

I Go to Sea

S/V *Moshulu*, East Side,
York Dock, Belfast

29 September 1938*

Dear Mummy and Daddy,

. . . I will try and give you some idea of what life is
like on board a sailing ship in port, from the point of view of
a humble apprentice.

At 5.30 in the morning the boy who is night watchman
kicks open the fo'c'sle door, shouting '*Resa upp, resa upp!*' so
that you think the Last Judgement has come. If you don't
'*resa upp*' he rolls you backwards and forwards as if you
were a pudding. This sometimes leads to a fight breaking
out, as there are some very quick-tempered people on
board. Eventually I climb out of my coffin – they are about
seven feet long and about two feet high and the crew fit
them up with various mementoes and curiosities to make
them more homelike, such as assegais [the ship had been to
Portuguese East Africa on the last voyage], horrible-
looking souvenirs made from Australian wattle wood,
pictures of girls, mostly film actresses. I have put up a
picture of a very luscious-looking girl from *Esquire
Magazine* and pretend she is my girl-friend which, almost
unbelievably, everyone believes. I have to make as brave a
show as possible being the only Englishman on board,
although there is supposed to be an American coming,
which would be nice.

Then after drinking coffee, quite good, with condensed
milk, and eating bread baked on board – very good –
having forced one's way into damp dungarees and boots,
we all go out on to the deck which is slimy and dirty, where
we wait with the boys from the other fo'c'sles for one of the
mates to tell us what to do.

The first two mornings I cleaned the lavatories. They
have no running water even if there ever was any, so

* This was the day on which, at Munich, Chamberlain, Daladier, Hitler and
Mussolini agreed to transfer the Sudetenland to Germany and to guarantee
the remaining frontiers of Czechoslovakia. Germany was now the dominant
power in Europe.

you have to haul up water from the dock and then use an iron rod to ram up and down them. The only way to use these lavs is to stand on the seat, which you are not supposed to do; but you have to. I had a fight because of this – 'Take dose bloddy boots off,' someone said when they found me standing on one of them. The doors have no locks on them, just like school. I wish Daddy had not insisted on my being so regular in my habits.

The second morning I and three others carried coal from what is known as the 'between-deck', an underdeck above the hold, to the cook's galley. The between-deck is rather dangerous because it is very dark and has openings in it below the hatches, leading down into the hold, and it is easy to fall down them. I haven't, as you can see from this letter. We got very dirty shovelling and carrying coal.

After that, each day, I was put over the side on a platform to do *knacka rost*, which is chipping rust from the ship's side with a hammer. It is a rotten job when it is raining and on the first day I dropped the hammer in the dock, which comes off my ten shillings a month wages.

Today I did all the washing up for three fo'c'sles, about twenty-eight men. It was a pretty revolting job as there were lots of uneaten sausages, uneaten because they weren't much good, all stuck in solidified grease, and the tables were all covered with jam and cigarette ends. What washing-up water you can get comes from the *kock*, the cook, who is pretty bad-tempered. They give you sand for scouring. I'm going to buy some Vim as everything has to be spotless after washing up and the boys create hell if it isn't. I wonder how my hero, T. E. Lawrence, would have got on here. He made an awful fuss about being an aircraftsman.

All my love,
Eric

I Go to Sea

Dear Mummy and Daddy,

. . . Last night I was night watchman. All the rest of
the crew went ashore, some of them to a dance hall in
Corporation Street where I went on the first night and
danced until someone started a fight and the police came. I
was given a pick helve as a weapon, and lantern, and told
to patrol the deck and not let anyone on board, except the
crew. Nightwatchman is a good job as the steward leaves
you out lots of bread and margarine and Dutch cheese, the
round red sort that tastes a bit like soap, and mustard
pickle, and you can make as much tea as you like in the
galley. Of course, with my being night watchman, a
lunatic, in which Belfast abounds, at least in the docks, had
to try to get on board and convert me to some weird
religion. By the time I had got rid of him the crew were
coming back and I had to carry out another of my duties,
which is to see that anyone who wants to be sick is sick over
the side, anywhere rather than on the captain's sacred
bridge-deck or outside his quarters where, unfortunately,
the gangway to the dockside is rigged. Nevertheless, in
spite of my efforts, one Finn managed to wrench open the
skylight of the midship's fo'c'sle and be ill through it over
the occupants, all of them senior members of the crew,
what are called 'daymen' because they don't normally
stand watches at sea, such as the assistant sailmakers, the
carpenter, blacksmith and so on, who were seated below.
For this I was blamed, which seemed unjust, but this is not
a just life.

The rest of the night was very long, until I woke the crew
with lots of '*resa upps*' and coffee at 5.30. On deck I listened
to the wind droning in the rigging and wondered what it
will be like at sea. How big the ship is. The lower yards
on which the biggest square sails will be bent are ninety-
five feet long and weigh each more than five tons. The
sails for them, there are three altogether, each weighs

99

two and a half tons, much more when they're wet. Altogether there are eighteen square sails and seventeen fore-and-aft sails, about forty-five thousand square feet in all and there are about three hundred ropes – braces, halliards, buntlines, clew lines, downhauls and so on – to control them; and I have to learn the names of each one of them in Swedish and where to find each one in the dark.

I was told all this by the sailmaker, who is also the bosun. He has been at sea for forty-three years, all of them in sail, and was so long in Scottish ships that he speaks English with a Clydeside accent.

Because I was night watchman I have the day off and am now eating an enormous high tea in this café while the rest of the crew are over the side, chipping. Soon we shall have to clean the bilges. Then we are going to a wharf where we shall take on board the ballast for the voyage out. 'Bloddy job' everyone says this is.

Yesterday, when I was very dirty, Ivy Anderson arrived in an open SS100, looking wonderful, all dressed in black and in a huge hat not really suitable for fast motoring. Everyone thought she was my girl-friend, another one besides the one I cut out of *Esquire*. I said she was an aunt but no one believed this, so I just let them think what they want to. The second mate fell for her to such an extent that he's allowing me to go and stay the night at her place up the Lough. She said she's asking Mummy to come over and see the ship off. I don't think this is a good idea. Do you? Too weepy. It looks as if there's not going to be a war.

All my love and thanks very much for the money. I'm going to buy lots of jam.

Eric

The next letter is from my father and is dated 6 October 1938. By this time the threat of war had receded. Czechoslovakia had been sold up the river, while four days later Germany occupied the Sudetenland in accordance with the Munich agreement.

My dear Eric,

. . . We are constantly thinking of you and wondering what new experiences you are having. You ought to have a bottle or two of lime juice as a preventative against scurvy. Slit trenches have been dug everywhere even, I regret to say, in the Royal Parks and in the St Paul's School playing-fields. They are proving extremely dangerous and a number of persons have already injured themselves by falling into them. We can only hope that they continue to serve no useful purpose. Duff Cooper has resigned from his post as First Lord of the Admiralty. He is a man of honour and was right to do so. Your mother who, as you know, writes constantly to you, sends all her love. She is being very brave about your becoming a sailor. I am glad you have been joined by a nice American fellow. I was afraid you might be the only real English speaker on board.

Steer clear of mean night adventures in the streets and God bless you.

S/V *Moshulu,*
Belfast

11 October 1938

Dear Mummy and Daddy,

We are now loading ballast, mostly rock and sand used in blast foundries, and paving stones and the best part of an old house. Altogether about fifteen hundred tons. There is bad news about George White, the nice American boy. He fell into the hold through the tonnage opening in the between-decks below Number Three Hatch when there was no ballast in it and broke one of his legs and various other things, so he won't be able to sail with us. I've just spent a wonderful time with Ivy and her husband at their house. The butler, called Taggart, brought me an enormous breakfast in bed and all my clothes were laundered for me, goodness knows how in the time.

My other friends are Jansson, the boy who put my trunk on board, a Dutch boy, called Kroner, and a Lithuanian named Vytautas Bagdanavicius, who sailed on the previous voyage. We go to gruesome pubs and drink porter, a weak version of Guinness and we eat fish and chips and go to the Salvation Army Hostel for wonderful hot baths. Don't worry about mean night adventures in the streets. They are much too wet, and we haven't got any money. The only one who tried a Matros, an AB [able-bodied seaman] got a nasty disease which he is treating himself, with a syringe. I wish he was elsewhere . . .

All my love,
Eric

This letter is from my father and is dated 16 October 1938. It was sent to Belfast but, having arrived too late, was forwarded to Australia.

My dear Eric,

I feel that you are on the eve of your departure for the open sea, and so I take leave to bid you a fond farewell. You have chosen a difficult job and are beginning life again on the bottom rung.

I enclose a German Text Book for Travellers which may help you with some words you have forgotten. I could not get one in Swedish or Finnish and I did not think Norwegian would be much good.

That fellow with the venereal disease sounds a rotten blighter. I should complain to the captain about him . . .

Good luck to you, my dear boy, and a safe journey.

Your loving father.

PS. Your mother is writing separately.

We sailed for Australia on 10 October 1938 with a crew of twenty-eight.

4 January 1939 38°33′S 132°21′E (78 days out) Great Australian Bight. Made 198 miles.
Someone put a lump of shit in my bunk. Thought it was Hermansonn who has been an absolute bastard for a long time. So when dinner came I picked up a bowl of custard and threw it in his face. Then we went on deck and fought.

It took some time but I won. The captain stopped the fight. It turned out that it was Holmberg who did it and I wanted to fight him too but I was dissuaded. Anyway, it didn't matter, the important thing was to wallop someone. The next time I'd do it the first day on board.

7 January   35°09′S 135°31′E (81 days out).
Raised Cape Catastrophe in the morning, on the west side of Spencer Gulf [named by the explorer Flinders in memory of a boat crew, lost there in 1802]. From it scrub-covered cliffs ran away north-westwards with rollers steaming in at the foot of them – a bloody spot. All day with the temperature up in the hundreds we beat about south of the entrance of the Gulf trying to weather the Cape, tacking ship, slithering on the decks which we have linseed oiled, like a lot of frying flying-fish, listening to the second mate's wireless as it literally poured out news, the first we have had for eighty-one days, from Adelaide, like a fruit machine with something wrong with it pouring out money – awful news of bush fires, frontier incidents in Poland, but at least of a world apparently not at war.

8 January   (82 days out).
A fair wind takes us in past the South Neptune Islands. Anchor off Boston Island 2.30 p.m., eight miles off Port Lincoln, having sailed about fifteen thousand sea miles. Already at anchorage four other Erikson barques, *Pommern*, *Passat*, *Viking* and *Pamir*. *Lawhill* arrived the following day. Only *Pommern*, a bald-headed barque (without royals) but one of the finest, strongest vessels in the fleet, has done better than *Moshulu*, seventy-eight days from Belfast.

We were three months in Australia. At first sweltering at anchorage waiting for a freight to be fixed, so that we could load a cargo of grain somewhere in Spencer Gulf, which runs up into the heart of the wheat belt, then, when hope had almost been given up of fixing a freight for any of the ships, and we had visions of sailing home in ballast or being sold with the ships like a lot of slaves, all the ships got freights and *Moshulu* was ordered to load a cargo at Port Victoria on the other side of the gulf at £1.37½ ($6.34) a ton – in 1938 she had loaded nearly five thousand tons at £2.06 ($10.30) a ton. The Spencer Gulf was a hell of a place, wherever you were in it in summer, plagued by flies and an appalling wind as hot as a blast furnace which poured down through it from the deserts of the interior, causing *Moshulu* and other ships to drag their anchors. To go ashore, we rowed and sailed eight miles to Port Lincoln and eight miles back. I found a lot of letters waiting for me and I sent my parents a telegram which read 'Muscular, happy, penniless' and got some money by return.

We sailed from Port Victoria at 6.30 a.m. on 11 March 1939, our destination Queenstown in southern Ireland (now Cobh), for orders, by way of Cape Horn. We were, in fact, taking part in the 1939 sailings of what was known as the Grain Race. This year turned out to be the Last Grain Race, and I wrote about it in a book of the same name.

We rounded Cape Horn on 10 April, Easter Monday, having sailed more than six thousand miles, and were fifty-five days to the Line. We were one day ahead of *Parma*'s record-breaking passage of eighty-three days from Port Victoria to Falmouth in 1933. That year she had been thirty days to the Horn, twenty-five to the Line, but in 31°N 47°W our luck deserted us and we failed to pick up the strong westerlies we needed to beat her.

On 9 June at 8 p.m., ninety days out from Port Victoria, we raised the Fastnet Rock, fifteen miles to the north-east. We had smelt the land for days. The following morning and until evening we were becalmed near the Rock. Five men rowed out to us from Crookhaven, near Mizen Head, nine miles. The captain made them drunk on rum and we left them drifting into the sunset in the direction of the New World. (More than twenty-five years later I got drunk in Crookhaven with the survivors of this long row.) Then a breeze came up and took us ghosting along the coast of Southern Ireland, past Cape Clear. Nothing could have been more beautiful to us than this country at this moment.

The following day at 5 a.m., the wind shifted from NW

through W to WSW, the best sort of wind and we squared away for Queenstown. At about eleven o'clock we took a pilot from a black and white cutter, heaving-to for him to come across to us in a rowing-boat. Then both watches went to the fore braces, boarded the fore tack and began to clew up the remaining course sails, before racing aloft to see which watch could be the first to furl the Main and Mizzen. We won, in the port watch.

We came to anchorage off the narrow entrance to Queenstown under a couple of topsails and staysails. It was twelve o'clock ship's time on Saturday, 10 June 1939, and we were ninety-one days out from Port Victoria, having sailed more than fifteen thousand miles.

The Pilot told us that we were first home, and although we did not know it at the time, we had won the Last Grain Race.

12

Snakes and Ladders
(1939–42)

When war broke out in the autumn of 1939 it proved remarkably difficult to join the armed forces. All the Royal Navy could offer me was the possibility of signing on for eight years as a rating, so I went to work on a farm near Salcombe in Devonshire and looked after an enormous army of pigs, until one by one all the girls went off to the war in various capacities and a dreadful air of sadness descended on the place. It was as if all the people I had known in that wonderful summer before the war were dead or had never existed. So I returned to London and eventually succeeded in joining the London Scottish, a territorial volunteer regiment affiliated to the Gordon Highlanders, as a private soldier. In order to be accepted it was necessary to prove that at least one parent was of Scottish descent, and this was how my mother, who was born of Devonian stock in Pimlico, London, came to be born, on paper, in Tobermory, Mull.

In the summer of 1940, with the French military alliance palpably collapsing, I was sent much against my inclination (for I enjoyed being a private soldier), to the Royal Military College, Sandhurst, to be trained as an officer. The RMC had only recently become OCTU, an Officer Cadet Training Unit, and the non-commissioned staff referred meaningfully on every possible occasion to a golden age 'when the gennulmen cadets were 'ere', although the non-commissioned staff still continued to refer to us as 'gennulmen', which was rather confusing. At the conclusion of the course I found that I had passed out with an 'A'.

It was a help that before the final written examinations one cadet had succeeded in obtaining printer's proofs of the papers, which he generously disseminated among the rest of the company; we all contributed to the purchase price! Perhaps the purchase was connived at. Whether it was or not, our officers, warrant officers and non-commission officers on the teaching staff all basked in the reflected glory of a company that had done so well, so everyone was satisfied.

In the autumn of 1941 I arrived in the Middle East from India. There I joined the Special Boat Section, whose job it was to land from submarines on hostile coasts in order to carry out acts of sabotage against railway systems, attack enemy airfields, and put ashore and take off secret agents. Members of the SBS had also made sorties into enemy harbours with the intention of sticking limpets (magnetic mines) on ships at anchor and blowing them up.

My interview for this job took place on board HMS *Medway* in the harbour at Alexandria, the depot ship of the First Submarine Flotilla which provided the SBS with transportation to its target areas. The interviewer was Roger Courtney, the founder of the Special Boat Section, an astonishing officer who had been a white hunter in East Africa and had canoed down the Nile. Over his desk was displayed a notice which read ARE YOU TOUGH? IF SO GET OUT. I NEED BUGGERS WITH INTELLIGENCE. This notice made me fear that I would not be accepted, but I was.

I spent the next few weeks at the Combined Operations Centre at Kabrit on the Suez Canal, learning to handle folboats and explosives, how to sink shipping, and how to blow up aircraft and trains or otherwise render them inoperative. Learning to sink ships involved swimming at night in the Bitter Lakes – which lived up to their name in the depths of winter – covered with grease and wearing long woollen naval issue underwear, and pushing a limpet towards whichever merchant ship lying at anchor had been chosen as the target, the limpet being supported by an inflated car inner tube with a net inside it. This was Britain's primitive equivalent to the highly sophisticated two-man submarines of the Italian Tenth Light Flotilla, which in December that year succeeded in entering the harbour at Alexandria and exploding charges under the battleships *Queen Elizabeth* and *Valiant*. Both ships were disabled and put out of action for months, seriously affecting the balance of sea power in the Mediterranean.

On my first practice attempt I was sent to attack a Dutch merchant ship at anchor in one of the Bitter Lakes. I reached it, thinking myself undetected in brilliant moonlight but wondering how anyone on board could possibly fail to hear the noise made by my chattering teeth. To set the limpet in its correct position on the ship's side it was necessary to dive deeply and as I did so I found myself enveloped in the contents of an entire Dutch lavatory pan which someone with a grotesque sense of fun and a remarkable sense of timing had released by pulling the chain.

'Better luck next time, *mynheer*,' a voice from the deck said as I came up spluttering. 'I should joose a dark night, if I was you, and dry not to make so much noise, even if it is so cold.'

Next door to our camp at Kabrit was David Stirling with his SAS, Special Air Service. As his success and power increased, David sometimes gave the impression that he was contemplating a takeover bid for SBS. We used to make use of some of his training facilities – he had a testy genius in charge of his explosives department, a Royal Engineer called Bill Cumper. He also had a lofty tower from which embryonic parachutists were expected to launch themselves parachuteless, and they also had to jump off the back of trucks going at about thirty miles an hour. His camp was definitely no place for the chicken-hearted. There was also a band of anarchists from Barcelona whom no one knew what to do with. They had murdered so many Egyptian taximen and buried them in the sand, instead of paying their fares like any normal persons, that it was now almost impossible to get a taxi from Kabrit to Ismailia and back during the hours of darkness, which was a bore.

One day, having climbed this tower to admire the extensive view and counting myself lucky that I was not called upon to jump from it, I was about to descend by the way I had come when I heard the voice of one of David's sergeants from far below say in Caterham* accents, 'No officer or man, sir, who has ever climbed that tower, has ever walked down the stairs! Once up there you have to jump, sir!'

* Caterham in Surrey was a training depot of the Foot Guards, where the characteristic high-pitched word of command peculiar to these regiments was taught, practiced and perfected.

So I jumped. I had no choice, and because I had not learned the basic facts about parachuting (which was not a condition of membership of the SBS), I hurt myself.

That winter, as part of a detachment of SBS, I was sent to Tobruk to operate with a flotilla of motor-torpedo-boats. Tobruk was extremely noisy. The Germans were intent on rendering it unusable as a supply port for the Eighth Army, and each afternoon Stukas, sometimes in large numbers, would come screaming down out of the sun with their sirens going full blast and with everything in and around the harbour firing flat out at them. In these moments, aboard one of the American-built MTBs which were loaded to the brim with three thousand gallons of 100 octane fuel and 21″ torpedoes, or on board the flotilla's minute depot ship, one felt curiously exposed.

13

Love Among the Ruins
(1942)

While we were in Tobruk four of us received orders to report without delay to the Directorate of Combined Operations at GHQ in Cairo; its Director was an excellent sailor, popular with the SBS, named Admiral Maund. We reached it late in the afternoon of the day we set off from Tobruk, after a hair-raising drive of more than four hundred miles down the desert road, or what was left of it, in a thirty hundredweight Bedford truck. We then went off to eat ice-creams at Groppi's Café before reporting to the DCO, in case we were being sent back to Europe in one of the high-flying Liberators (which at that time provided a regular service between Britain and the Middle East), to sink the *Prinz Eugen* with our limpet mines and would not have another chance.

At the DCO we were only asked if we possessed prismatic compasses, and when we said that we did we were told to report to the Naval Officer in Charge, NOIC, at Beirut without delay. Before the shops shut in Cairo that evening I bought two guide books, one to Palestine, the other to Syria and the Lebanon, and *The Quest for Corvo*, by A. J. A. Symonds.

We drove all night in turns – it was a close fit for all four of us in the cab of the truck – stopping to doze uneasily for a couple of hours during the interminable crossing of the Sinai Desert, which was even more perishingly cold at night than the Western Desert. Early on the morning of the second day out from Tobruk, we arrived at Gaza: we had covered about eight hundred miles since setting out from Tobruk, which was not bad, considering the terrible state of the desert road.

Then we were in Palestine and suddenly it was spring: the countryside in the coastal plain was a green paradise with burgeoning fields of wheat and barley; meadows and gentle hillsides were scattered with wild flowers; and everywhere there were groves of lemons and oranges, of which we bought whole box-loads, afraid that this other Eden might be only a temporary phenomenon and might be succeeded by yet another dusty wilderness as we drove northwards. After the Western Desert, where it was still winter, after Sinai, where we had lain shivering in the cold grey hours before the dawn in the duffel coats we had wangled from the navy, to be in Palestine was to be born again. Armed with the guide book to Palestine, I was able to persuade my companions to make a number of short detours from the main road in order to visit places, some of which I had learned about in Divinity at school, none of which I had ever expected to set eyes on. Thus I saw Askalon of the Philistines, whose goddess was the mermaid, Ashtoreth. Conquered and despoiled innumerable times by Egyptians, Assyrians, Jews, Macedonians, Persians and Romans, re-edified by Herod who was born there, the city was finally demolished by Crusaders and Saracens. By the time we arrived, ancient Askalon was nothing but some heaps of stones and broken pillars scattered among the mimosa and groves of pine and citrus on a cliff, below which the Mediterranean crashed on a then deserted shore.

Next door to it was Asdod, another of the five cities of the Philistines (the others were Gaza, Gath and Ekron), where they had worshipped Dagon, a god of fertility in the form of a merman, an idol with the head of a man and the tail of a fish. The Egyptians had besieged it unsuccessfully for twenty-nine years. Now there was even less of ancient Asdod than there was of ancient Askalon.

More extensive were the ruins of Caesarea Palestinae,* established by King Herod to be the seaport of his inland capital, Samaria. By this time, as a result of haggling for boxes of oranges, shopping in Tel Aviv, which I thought had a distinctly Eastern European air about it (although I had never been to Eastern Europe), visiting ruins, in fact generally behaving in a

---

*It was called Caesarea Palestinae to distinguish it from Caesarea Philippi (now the village of Banias), in south-west Syria near the Israeli frontier, where the springs of the Jordan rise; the Caesarea that is now Cherchel, a small town between the Atlas Mountains and the sea, west of Algiers; Caesarea Mazaca (now Kayseri), in Turkey, ancient capital of Cappadocia; and others.

thoroughly unmilitary way, some of the impetus had gone out of our expedition.

When we reached Caesarea the sun was setting, drenching everything, including the remains of a Roman aqueduct which stood among the dunes that had engulfed what remained of the ancient city, in a brilliant ochrous yellow light. Waves were breaking over the remnants of the ancient harbour works, and down on the foreshore where the air was full of flying spume, a man on a camel, dressed in rags which were streaming in the wind, was the only other human being in sight. It was here that I began rooting about underfoot with a stick, turning up potsherds and iridescent fragments of what may have been Roman or Byzantine glass. In the meantime, whoever was driving kept the motor running and a brother officer shouted as he had done all day whenever I had found a ruin to my taste, 'For Christ's sake get a move on, Eric, we can't stay here all bloody night!' while the sergeant cried rather more plaintively, 'Oh, do come on, sir!'

That night we slept at Zikhron Ya'kov, a village on the southern flanks of Mount Hermon. It was surrounded by vineyards planted by Baron Edmond de Rothschild and was what I imagined, aided by memories of photographs in *The Children's Colour Book of Lands and Peoples*, a Central European Jewish village might have been like before the war. That evening, sitting there in the twilight with the bats flittering overhead, drinking the excellent red wine, and eating the flat bread and a substantial soup called *kreplach* made with chopped meat and dumplings, the Western Desert, with its graveyards of bombed, shelled and blown up vehicles, and its huge, lethal minefields, seemed far away.

The next day, the third since leaving Tobruk, we drove on northwards. At Acre we looked down into a dungeon in which the despot Jezzar Pasha, who had successfully defended the place against Napoleon, used to pelt his prisoners with cannon balls through a hole in the ceiling. It was a tough campaign. In the course of it Napoleon ordered the mass executions of prisoners by firing squads. As we flashed by on what had been the Roman road to Syria we saw milestones, inscriptions recording innumerable wars and conquests, fragments of altars, gaping catacombs, plundered sarcophagi and other ruins of the past.

One of these remains, at which I was allowed a brief stop, was a sarcophagus said to have contained the remains of King Hiram of Tyre, a splendid Phoenician tomb chest with a pyramidical cover, set on a pediment ten feet high. Another even more fleeting halt was made for my benefit at Alexandroscena, Alexander's tent, where the Macedonian camped while conducting the siege of Tyre. There was no tent any more, only a now disused *caravanserai* for the reception of travellers, their animals and goods. The only visible remain of ancient Alexandroscena was a spring gushing in a rock basin. Yet it was a magic place, and one at which I would have dearly liked to linger.

Tyre, captured by Alexander after a seven-month siege, was another truly ruinous ruin. As was Caesarea, most of Tyre was covered with drifts of sand, and the most visible remains were numbers of huge pillars lying where they had either been thrown down or collapsed of their own accord. Even comparatively modern remnants of Tyre were scanty. Once there was a great basilica, built by the Venetians on the site of an earlier one and supposed to contain the remains of Frederick Barbarossa, who was drowned in 1190 at the mouth of a river on the coast of southern Turkey while on his way to the Third Crusade. Now all that remained was a crumbling wall on some waste ground on the outskirts of the town, used by some local inhabitants and a horde of mangy dogs as a convenient place to relieve themselves. Most of its stones, as those of Caesarea and Askalon, had been carted away to build Acre.

Ancient Sidon, older than ancient Tyre, was even more ruined, in the sense that it was even more effectively buried beneath Crusader, Muslim and other remains. Its streets were narrow trenches spanned by arches and full of Arabic-speaking Jews, but it was surrounded by delightful gardens in which oranges, lemons, apricots, medlars and almonds all came to maturity in due season. Offshore on an island was a ruined Crusader castle, the Castle of the Sea.

'Oh, do get a move on!' the others said, or words to that effect. There was no time to see the necropolis of ancient Sidon, no time for a detour to see the monastery which had been the home of Lady Hester Stanhope and the place where she was buried. In less than an hour we reached Beirut and, together with other SBS who had already arrived, received from the NOIC the details of Operation Aluite.

To us Operation Aluite seemed pretty defeatist. It implied that a German advance into Syria through Turkey, a recurrent British nightmare, one brought on by the continued presence of von Papen as German ambassador at Ankara, would be followed inevitably by the Allied evacuation of Syria and the Lebanon, probably that of Palestine, and eventually, by implication, that of Egypt.

In an endeavour to stem such an invasion a great fortress was being built near the port of Tripoli, north of Beirut, where the northern arm of the Iraqi pipeline came in. In the event of such an invasion being successful, it would be the task of the SBS, working from Cyprus, if Cyprus had not itself fallen, to act as guerillas and sabotage the German lines of communication. In anticipation of such a disaster, we were to sound every inlet in which a clandestine landing could be made, to map the hinterland leading up to the main road and the railway, and to make an assessment of every bridge and other important installations with a view to blowing them up, on the two hundred and fifty miles of coast between the Turkish-Syrian frontier and the Lebanese-Palestinian frontier, north of Haifa. We were also to seek out suitable hiding places for caches of explosive and ammunition. All this had to be done without the knowledge of our French allies in these parts, which seemed impossible.

In order to assist in the concealment of such dumps, an enormous quantity of artificial, lightweight rock of the same colour and texture as that found on the coasts of Syria and the Lebanon had been manufactured from papier-mâché. What conclusion a wandering goatherd or even a German *feldwebel* would come to when he found himself walking on artificial rocks made from papier-mâché will never be known as they were never put to the test.

'Personally, I think the whole thing's rather a waste of time,' the NOIC said, 'although you should have fun. Nevertheless I don't envy you. As you know, the Free French have only recently taken over the country from Vichy and some of the permanent members of the administration are thoroughly untrustworthy and hostile and loyal to what they call "La France de la Metropole". The coastguards are said to be particularly trigger happy. If they shoot at you I should shoot back and ask questions afterwards. Oh, and take plenty of rubbers and pencils.'

That evening, together with another officer from SBS, I put up at the St George Hotel on the waterfront where we met a couple of extremely forward Greek girls who were living there. They told us that they had managed to escape from Athens when the Germans moved in and had ended up at Beirut. Soon we were on terms of some intimacy with them and we arranged to see them again when we returned to the hotel at the end of the following week.

The following morning the four of us, including the sergeant who had said 'Oh, do come on, sir!' in such a pained voice, now himself transformed into an enthusiastic ruin-fancier, left Beirut in the thirty hundredweight truck to start work on mapping the section of coast south of the Turkish frontier, while others began on the coast south of Beirut. We had been given enough money to enable us to live off the country without having recourse to other military organizations, so much in fact that we decided to take with us a Sten gun as well as pistols, just in case there were robbers about.

All that day we drove north, crossing the Nahr el Kelb, the Dog River, guarded in ancient times by a savage dog with a bark so loud that it could be heard six miles off. In its gorge, where it entered the sea, we saw inscribed slabs recording the passage of Egyptian Pharaohs, Assyrians, Babylonians, Greeks and Romans, the Emperor Caracalla and the Third Gallic Legion, British and French troops in July 1918, and a triumphant one recording the entry of the French into Damascus in July 1920 when they expelled King Feisal and the Arabs. Seeing these great *steles* I realized that we were just another band of marauding soldiers. We saw the ruins of ancient Byblos, the principal city of the Giblites, claimed by an ancient Greek, Philo, to be the oldest town in the world, with a square Crusader castle above it on a hill, ruins among which we later lived for some days while making our survey.

On the outskirts of Tripoli work was proceeding on the construction of the fortress, which looked pretty feeble to us, and all along the coast hordes of men were working away on the Naquara-Beirut-Tripoli railway which we were already making plans to destroy.

Early in the afternoon, four and a half days from Tobruk, we reached the outskirts of Latakia. By now all of us were suffering from an accumulation of fatigue. There, travelling at nearly sixty miles an hour, whoever was driving encountered a horse-drawn vehicle lumbering towards us on the crown of the road which he would have almost certainly have been able to avoid if whoever was sitting next to him had not momentarily lost his head and grabbed the wheel. As a result we crashed clean through the parapet of a bridge – which, fortunately as it turned out, had the effect of slowing us up – and down into what turned out to be, equally fortunately perhaps, the dried-up bed of a river, where the vehicle performed a somersault before coming to rest lying on its left side with the only available exit door on the driver's side jammed tight, its engine still running merrily, and its windscreen unbroken.

At this moment none of us was sufficiently in command of his faculties to switch the engine off but fortunately, before the truck had time to burst into flames, we were rescued by members of an Australian gunner unit who had witnessed the accident. They broke the windscreen and one of them switched off the engine with the words, 'You jokers ought to have a sign out, "Frying tonight"!'

By some miracle no limbs were severed, no blood flowed, and none of us suffered anything but severe shock, probably because the four of us were such a tight fit in the cab; although the other three, one of whom weighed more than sixteen stone, fell on me when the truck landed on its side.

The Australians took us back to their camp and fed us and gave us beer after their medical officer had pronouced that we were relatively unscathed. Then, when we were all safely tucked up in one of their tents, they salvaged our truck which they immediately proceeded to cannibalize, being desperately short of spare parts for their own transport in this remote spot, so that when, the following morning, we tottered over to eat breakfast with them and asked how our truck was we were met with a lot of level gazes and the words,'Which truck, sports?'

It was difficult to argue with people who the previous day had saved our lives and who looked perfectly capable of cannibalizing us and burying us in one of their gun-pits if we continued to argue the toss with them. So, after procuring a document from their transport officer testifying that our truck had gone up in flames and had become a total loss, we thanked them for their hospitality, stole a pair of their binoculars from the tent in which we had passed the night, and accepted a lift from them into Latakia.

We spent the next couple of days listlessly lying around in a rather decrepit casino which enjoyed fine views over the sea, most of the time trying hard not to burst into tears – we really had been badly shocked. Fortunately there were no croupiers, otherwise it might have been difficult to resist the temptation to use some of the imprest money we had been given for Operation Aluite to play at one of the tables. After this we began to feel better and started work.

We then hired from an Armenian entrepreneur a big, beat-up, black American automobile in which, armed as we were, we looked rather as if we were setting off to massacre the Moran Mob in Chicago on St Valentine's Day, 1929.

The next weeks were the best that any of us had so far experienced in the course of the war, and the best that many of us were ever likely to experience: swimming about in lonely coves, taking soundings with long canes cut in some convenient plantation; pacing out base lines and using our compasses to make triangulations across the fields of wheat and barley; searching out caves and rock tombs, in which the coast abounded, that might serve as caches for explosives; swinging about like apes among the girders of railway bridges; meeting primitive-looking goatherds, one of whom, a rather elderly Sunnite Muslim I found sitting on a rock, had been an itinerant pedlar in the United States before returning home to marry a Syrian girl, a marriage which he said he had arranged by post.

In the course of our travels we encountered a remarkable

diversity of religions and nationalities. There were Alawites, Druzes and Ismailites, whose religions contained elements of Muslim, Christian, Indian and Persian beliefs. There were Armenians who had either been deported from Turkey during the First World War by the Turks or who had left it of their own volition to avoid being slaughtered – it was some Armenian Orthodox Gregorians who invited us to a village near the Turkish frontier for a play in Armenian lasting eight hours, which was a kind thought but a great trial to all of us. There were Armenian Catholics and Maronites and Greek-Catholic Melkites and Syrian-Catholic Syriacs and Chaldean Catholics and Roman Catholics and Greek Orthodox and Syrian Orthodox, which included Jacobites and Nestorians, and there were even some Protestants. There were Sephardic Jews and there were Sunni and Shiah Muslims, the latter so fanatical that if they were forced to feed an infidel they destroyed the crockery as soon as the visitor had departed. And in Syria there were some very odd people indeed, whom we never saw, called Yezidis, who believed that God had passed the administration of the world over to the Devil, which on second thoughts did not seem odd at all. None of these conflicting sects seemed particularly fond of one another.

We also occasionally met the dreaded coastguards, with whom we endeavoured to ingratiate ourselves when they looked like turning nasty, which they often did, by offering them cigarettes. Once we were shot at, as the NOIC had predicted we might be, by one who was a damn sight too loyal to la France de la Métropole, considering he had probably never been there. At night we either slept under mosquito nets beneath the stars, or among ruins, or very occasionally in rude, malodorous inns, the most rude and malodorous of which was called Le Grand Fleur de Tartousse built over one of the medieval sewers with which this famous town was riddled. Tartus was the last mainland stronghold of the Templars before they escaped to Cyprus, and their great fortress, built of enormous blocks, with the hall of the Order within it, still had the postern gate by which the knights made their escape to their ships when it was captured by the Mameluke Sultan Qualaun in 1291. There was also an empty, echoing cathedral, Our Lady of Tortosa, that had been constructed on the site of a chapel which housed an ikon of the Virgin. Together with the altar, the ikon survived a severe earthquake in 387, and thereafter both ikon and altar became objects of pilgrimage and veneration.

Offshore was the island of Ruad, the Phoenician Arvad. Half a mile long and a quarter broad it had a population of three thousand Sunnites, the women more heavily veiled than any others we saw on the entire coast. The men, who were all either sailors or shipwrights or both, wore striped shirts, huge baggy trousers and Phrygian caps. The walls of Ruad rose straight from the sea and tiny, conical windmills of whitewashed stone with canvas sails, which could be furled when not in use, stood among the rocks. In the harbour were caïques with mainsails set on sprits and with square topsails and lots of jibs set on enormously long bowsprits, and brigs and other rigs either forgotten or unknown in Europe. On the seaward side of the island were the shipyards to which trunks of trees were floated out from the mainland, where they were trimmed and shaped using methods that had probably not changed much since the ships of the Pharaoh Snefru visited these coasts around 3200 BC. Here I bought a beautiful model of a caïque, commissioned by some ambassador in Cairo who had not yet turned up to claim it.

One night we slept among the ruins of Marqab, one of the castles of the Knights Hospitallers. Built of black basalt and reached by a spiral track nearly four miles long, it occupied a fantastic situation, high on a spur of an extinct volcano. It looked impregnable but it had finally been captured by Qalaun in 1285 after a forty-day siege, although it had food supplies for five years; but even so it was a much better site for a fortress than the one that was being built down the hill in Tripoli, as good as Monte Cassino, and as difficult to take.

In its chapels, halls and passages and on its circuit walks, sheep and goats wandered. It was cool up there and we built a great fire in one of the rather smelly halls. We thought the flickering of the light on the walls and the giant shadows highly romantic until we were infested by bats, attracted by the flames, which were so numerous that they eventually forced us to retreat to one of the roofless towers. There we passed the rest of the night free of them and the rotten-chocolate smell of their excrement.

We used to start work as soon as it was light, then rest in the heat of the day. We had soon lost most of our external military characteristics. Down on the shore, under the mountains, it was much too hot to wear anything but shorts and sandals and straw hats, and the sergeant acquired a long, lean hunting dog. We were dressed like this, the sergeant with his dog on a rope leash and carrying the Sten gun, the others armed with pistols and the long canes we used for taking soundings, when one morning the Duke of Gloucester, on a tour of inspection with a convoy of military big-wigs, looked down incredulously on us from his motor car as he whizzed past, covering us with dust.

During the hot, midday hours we used to sprawl in the shelter of an ilex or an over-size boxwood tree in which the coast abounded, more often than not surrounded by ruins – we soon learned that the ancients had already identified for us all the best landing places, however insignificant. There we reclined drinking wine, eating chickens we had bought from some cook shop, carving up big loaves of bread that looked as if they might have been baked by Phoenicians, dreamily listening to the droning of unidentifiable insects, the shrill screaming of the cicadas or the endless din set up by the frogs in some nearby marsh, and sniffing the pungent smelly *maquis*.

We worked at Ibn Hani, a place lost among olive groves, where there were the remains of a temple and an amphitheatre, and at Ras Shamrah, the site of Ugarit, a famous city of the Phoenicians but with origins far older, going back to 5000 BC, perhaps further. Much of Ugarit was buried under Ras Shamrah, but a French expedition had continued to excavate it until the war put an end to their labours. Now it was completely unprotected – there were no custodians to harass us – and unvisited. In places one could look down, strata on strata, fifty feet or so, through different levels of civilization to where people had lived who had worshipped Baal, the God of Rain, and Dagon, to levels at which the inhabitants had had relations with Egypt and Crete in the seventeenth and eighteenth centuries before Christ; to where later, in the fifteenth and fourteenth centuries, they had installed sanitation and constructed burial vaults; and to the level where, in the thirteenth century, the Mycenaeans had lived in it until its final extinction by some peoples of the sea in the twelfth century. It was here at Ras Shamrah that, scrabbling among the rubble in the fearful afternoon heat with my stick, keeping watch while the others slept, I discovered two exquisite miniature bronze bulls which subsequently reached England, only to be stolen from over a fireplace in a drawing-room while we were having a Christmas party.

On Saturday mornings we used to return to Beirut in pursuit of pleasures to which the ancients themselves were no strangers. Together with the officer who, like me, had been attracted by the two Greek escapees from Athens, I used to put up at the St George Hotel, where they continued to stay. We saw nothing of the city. Once or twice we all four lunched together at a restaurant perched on the cliff overlooking some impressive offshore stacks called the Grottes des Pigeons. Another time we went to the mountains and stayed in a village. For the rest of the time we sunbathed with them on the hotel beach, swam, made love, drank, sunbathed, swam, made love, ordered up club sandwiches and so on until, at dawn on Monday morning, I used to board our sinister-looking motor car to cries of 'Oh, do hurry up, sir!' and roar away for Ras Shamrah, Tartus or wherever we happened to be making our maps. We were completely exhausted, unlike our partners who, unknown to us, had other bedfellows of a more senior, stay-at-home kind during the week. We had met our match, we both agreed, in these girls, who, for us at least, did everything for love.

Our only real problem during these wild, acrobatic weekends was what to do with the bulging haversacks which contained those fruits of our labours which we were still working on – those maps and reports that, once they were completed, were immediately despatched by way of the NOIC to Combined Operations Headquarters in Cairo, a world away. We resisted the suggestions of the manager of the hotel, always solicitous for our comfort, who saw us encumbered with them on the beach and at the bar, that we should entrust these haversacks to the hotel safe, in the same way as other guests no doubt had consigned their bibelots from Cartier & Boucheron before the war. Neither, however irresistible they were on other planes, did we have sufficient faith in our companions to put them in their hands, although they never asked us to do so. Their means of sustenance which allowed them to stay week after week at this expensive hotel was, to put it mildly, mysterious (although our suspicions proved to be unfounded). In the end the best, if not the only solution, seemed to be, apart from spending the weekends among the ruins of Ras Shamrah, to continue to do what we had been doing up to now – carry them with us always and when we were with the girls stow this material under them and the mattresses.

During this time, in the brief intervals when I had time to read anything, I read Symons on Baron Corvo, wishing that I could lay hands on the Baron's own book, *Hadrian the Seventh*, and also a lot of Hemingway, whose feeling for the sun-drenched open air and the pleasures of a physical existence seemed, almost uncannily, to complement the kind of life that we ourselves were living. Among them was *For Whom the Bell Tolls*, which had only recently been published. I gave it to the officer masquerading as a sergeant to read, and eventually almost everyone in the SBS read it. Up here in the Levant we could not hear the bell tolling; up here in the Levant in the spring of 1942 you would have needed an ear trumpet to hear it.

After this extraordinary, almost dream-like interlude in our military careers, we all returned to the fields of action from which that fickle goddess Fortune had fancifully removed us. However, before doing so, we delivered to the DCO by way of the NOIC the final instalments of our labours which, altogether, were of almost encyclopaedic proportions.

This mass of material, flavoured with a surprising amount of newly-acquired culture – the reports of the cultivated and gallant sergeant made particularly good reading – had an extremely short life. Consigned to the most secret archives, the whole lot was used a few months after we delivered the final sections to stoke the already huge funeral pyres of documents that were on no account to fall into enemy hands – although what use they would be to the Germans with Egypt already in their hands it was difficult to imagine – pyres which created a dense pall of smoke over Cairo in that summer of 1942, when it seemed more than probable that Rommel would arrive in the city in person.

14

A Trip to Italy
(1942)

Telegram – September 1942. War Office. Absolute Priority.

To G. A. Newby, 3 Castelnau Mansions, Barnes SW13.
Regret to inform you that your son Lieutenant G. E.
Newby is missing. Letter follows.

I was captured on the morning of 12 August 1942, together with
five other members of M. (Malta) Detachment of the Special
Boat Section, two or three miles off the east coast of Sicily. It was
a calm and beautiful morning. As the sun rose it shone on Mount
Etna visible in the haze to the north, a truncated cone trailing a
delicate plume of smoke. There was none of the vulgar colour
effects noted by Evelyn Waugh which had so upset him while on
a cruise in these same waters between the wars.

'Grouse shooting's beginning. Third year I've missed it,'
George Duncan, who was in charge of what was called
Operation Whynot, pointed out to us as we trod water, very
tired now as we had been in it for many hours.

We had been attempting to rejoin the submarine from which
we had landed to attack a German airfield, on which large
numbers of JU 88s had been assembled to destroy a convoy of
fourteen merchant ships, eleven British and three American.
This convoy was being fought through the Mediterranean at
fifteen knots from Gibraltar to Malta in order to save the island
from what would otherwise be inevitable capitulation.

Having sailed from Malta in *Una*, one of the smaller sorts of
British submarine, we made a successful landing on the coast
south of Catania on the night of 11 August in line with one of the
main runways of the airfield to which JU 88s were coming in to
land a couple of hundred feet overhead. The noise was
deafening. When we got close in we lowered ourselves into the

water from the three canoes and swam them in through the surf. There was no one to welcome us. We were in luck. We had landed midway between two concrete blockhouses, and if there were supposed to be sentries patrolling between them they must have been elsewhere. We carried the boats up the beach, buried them and obliterated our footsteps. The beach with the surf booming on it seemed at that moment the loneliest place in the world. Then we began to cut our way through the wire, praying that there were no mines underfoot and that the airfield when we reached it would not be alive with savage German police dogs, as it had been at Maleme on the north coast of Crete when the SBS had raided it unsuccessfully earlier that summer. For the first time I was in Europe. Twelve hours later what was left of the party was dragged from the sea some miles offshore and we were prisoners.

Of the fourteen merchant ships which took part in Operation Pedestal, five reached Malta, including the tanker *Ohio*, which was enough to save the island from surrender. The remaining nine were all sunk, four of them by JU 88s operating from Sicily on 12 and 13 August.

Having narrowly escaped being shot as saboteurs by a firing squad in the moat of a fort at Catania in Sicily, our little force was dispersed to various prison camps in Italy. Three of us were sent to Campo Di Concentramento PG21 at Chieti, a few miles inland from the shores of the Adriatic at Pescara.

The camp was already filled to the brim with officers who had been captured during Rommel's big offensive in June 1942, when he took Tobruk and drove the Eighth Army out of Cyrenaica. Before being sent to Italy in ships' holds they had suffered considerable privations and what the three of us now saw, guarded by Italians who were altogether too cock-a-hoop for our liking about the way the war was going, was a ragged band, many of whom we had known in the desert or else as elegant debonair figures on leave, propping up the bar at Shepheards in Cairo or eating *cailles-au-riz* in the Union Club at Alexandria. The first person I met when we arrived was a tough Welsh boy, a notable boxer, whom I had known at Colet Court and St Paul's, now a regular officer in the South Wales

Borderers, who later escaped, rejoined his regiment and was subsequently killed in action. His entire uniform consisted of a shirt, the remains of a pair of shorts and a pair of canvas shoes as full of holes as his trousers.

'Do you know,' he said, in a voice filled with awe, 'this place is filled with people from Radley' (an English public school on the banks of the Thames near Oxford). 'It's quite unbelievable but they've formed an Old Radleian Society and they all sit around talking about when they were at school together. There they are. Over there.' And sure enough seated on the ground in one corner of the compound there was a ragged little band of Old Radleians, talking about the past.

Several attempts at escape had already been made from this camp and I joined a small party which began work on a tunnel in a room used for peeling potatoes for the camp kitchens. While part of the party worked away on the potatoes, others worked on the shaft. On one occasion our look-outs failed to warn us that a Count de Salis had entered the camp on a visit of inspection on behalf of the Red Cross, attended by a phalanx of Italian functionaries which included the Italian Commandant and Capitano Croce, the camp interpreter, a repulsive Blackshirt who eventually came to a sticky and unlamented end. There was no time to replace the lid of the tunnel in the mouth of the shaft before this resplendent band swept into the peeling shed and while the good count asked us questions about our welfare one of the working party was forced to remain seated with his bottom in the hole still dementedly peeling potatoes as if his life depended on it, while the rest of us stood deferentially in an attempt to screen him from view, which the count must have found distinctly odd. This tunnel was subsequently discovered and a large collective fine was imposed on the inmates of the camp for the damage caused, as was the custom both in Italy and Germany. It was lucky that we were on the winning side otherwise we would all probably still be paying off these debts to the Axis.

The most original escape attempt was made by someone who managed to lower himself into the town sewers and paddle along them on an air bed. Unfortunately they were full of inflammable gas and he blew himself up while lighting his pipe and had to be given hospital treatment before serving a sentence in solitary confinement.

PG22 was the most cultivated camp I was ever in. It was more like a university than a prisoner-of-war camp. In it one could read philosophy with a tutor who had got a First in Greats at Oxford (the final examination for Honours in *Literae Humaniores*), study psychology with a resident psychologist, learn how to draft a foreign office despatch from a diplomat who had abandoned the service for the army for the duration of the war, listen to one of six orchestras, one of them a symphony orchestra of twenty players whose distinguished conductor composed his first symphony at Chieti before diving from a moving train in Italy later in the war in order to avoid being taken to Germany, after which, while free in the Italian countryside, he began collecting folk music. Almost every week there was a new play and some of the actors were almost permanently in drag – memorable was Bill Bowes, the enormous Yorkshire cricketer, in *Of Mice and Men*, brilliantly cast as Lennie, who would also teach one to play cricket; and when the first Americans arrived, taken in North Africa, they introduced us to baseball. Although I didn't realize it until much later, it was at Chieti that I first discovered my métier as a writer, if I can describe it as such, although nearly fifteen years were to pass before I was to realize it fully.

At the beginning of May 1943, after a truly fearful winter in which the camp was swept by infective jaundice, I was transferred with a number of other prisoners to a prison camp in what had been an orphanage on the outskirts of Fontanellato, a village in the Po Valley near Parma where we lived in what seemed to us such unimaginable luxury, sleeping in beds instead of bunks, sitting on chairs and dining off tables with cloths on them. Rumours began to circulate that we were being groomed for repatriation in some great mass exchange of prisoners, rumours that were entirely without foundation, although by this time the German Army had been driven out of North Africa. If Chieti had been a university Fontanellato was like a London club. The food was excellent. We handed in our Red Cross parcels to the mess and the whole lot was properly cooked and served, relieving us of the necessity to grovel on all fours blowing away at makeshift stoves constructed from old tin cans and burning our bunk boards as we had done at Chieti. There was even a bar which served truly awful wine bought on the black market, which flourished.

Extracts from a letter to Lieutenant Anthony Simpson, Royal Artillery.

PG49

20 June 1943

Dear Tony,

. . . There is an astonishing man here whose tank blew up in the desert – and no wonder – who lights his farts. He does it under cover of darkness, partly from *pudeur*, partly because the effect is more impressive. A long blue flame is emitted. I would not have believed it if I had not seen it with my own eyes. One wonders if, back in 'Civvy Street', he will be able to open safes with it . . . There is also a man who has become a millionaire by prison standards – the universal currency being cigarettes. He rescues teeth, watches and cigarette lighters that people let fall while using the Italian-style lavatories over which one crouches astride with one's feet on what look like engine-turned footplates until, without warning, an enormous head of water like the Severn Bore comes surging up from unimaginable depths and fills one's boots. Besides such heroic acts of salvage this man – whose firm he calls 'Finders' – acts as a broker or entrepreneur, putting people in touch with one another who want to swap, say, a silk scarf from the Burlington Arcade for a Viyella shirt, or a pair of corduroy trousers for a Dunhill pipe . . .

On 8 September an armistice was announced between the Italians and the Allies, the Germans took over the country and with German troops on the way to take over guard duties until we could be transported to Germany, the following day, we broke out of the place with the connivance of the Italian commandant, who was subsequently himself sent back as a prisoner and he received such rough treatment that he died. I myself left on a mule as I had broken my ankle falling downstairs

127

some days previously. The following day, while hiding in a hay loft, prostrated with hay fever, I met a very determined and personable Slovene girl whose father, a notable anti-Fascist, was the local schoolmaster. It was she who produced a doctor who had me transported to the maternity ward of the local hospital, where he put my ankle in plaster. While hidden in hospital I was visited by Wanda, who took time off from supplying other prisoners with food and civilian clothing to give me Italian lessons. When the Germans discovered that I was in the hospital it was she who gave me sufficient warning to escape, and after various adventures it was the Italian doctor who drove me down the Via Emilia, the main German line of communication with the battle front, to a place of relative safety in the high Apennines. Both the doctor and Wanda's father were subsequently arrested by the Gestapo but survived the war.

I was recaptured in January 1944 in a lonely hut in the mountains, having been betrayed, together with two other friends from PG49, by the local schoolmistress.

15

Conducted Tours with the Third Reich

(1944)

Northern Italy in January 1944 was gelid, frozen solid, without light, heat or hope, the Allies bogged down on the Sangro hundreds of miles to the south, with Monte Cassino yet to be stormed, Mussolini once more at liberty and the Fascists, now known as *Repubblichini*, once more in the ascendant, torturing and committing hideous crimes, aided and abetted by the Gestapo. The *Milizia*, the odious force recruited from the scourings of Italy, who had captured us, took us down icy mountain roads to Parma. By this time none of us was in particularly good shape.

Parma was like a city of the dead. There we were interrogated by a Fascist police chief in a fog-filled palazzo which, although extremely lugubrious, was at least warm. By this time my Italian was sufficiently voluble for me to be able to suggest to him that as his days as police chief were undoubtedly numbered it might be a good idea for him to let us go and thus cultivate a reputation for humanity which we would testify to when the Allies arrived.

'I quite agree,' he said frankly, 'about my time being limited. It is undoubtedly limited for all of us who refuse to betray our country to the British and the Americans and the forces of Communism, but I promise you we will continue to have a good time here until we are finished,' and he consigned us with a wolfish grin to the Cittadella for onward transmission to Germany.

The Cittadella, built in the sixteenth century by the Farnese, was a huge, star-shaped, brick fortress surrounded by a wide deep moat. We were accommodated in the Sala di Punizione which stood just inside the main gate. Balls and chains lying about in a wired-in compound outside it were presumably used for taming recalcitrant prisoners. An upper room, to which we were taken, had been heavily decorated with graffiti by previous occupants of various nationalities, some of whom had been betrayed in somewhat different circumstances from those in which we had been denounced, if what was scribbled and scratched on the walls was to be relied on – 'Oh, Mima, you beautiful fucking cow,' one English soldier wrote, with a mixture of lust and despair.

However, there was not much time for reading graffiti. The room was already occupied by a number of officers and men of the Greek Army, who had also been on the run. They were as intent as we were on taking what would probably be the last chance to escape any of us was likely to have before we were sent to Germany. They had already made a start, digging a hole through the ceiling, and we spent the rest of the night, much of it on one another's shoulders, trying to break through it on to the roof from which we hoped to be able to clamber down into the moat. Unfortunately the roof had been built to be proof against seventeenth-century siege mortars and in spite of all our efforts when the time came, six o'clock on a freezing morning, to be taken to the railway station, although we had made considerable progress and a terrible mess – we had hidden the debris under some broken-down beds – we still had a long way to go. We left the Cittadella with despair in our hearts.

However, we did not go directly to Germany. Instead we were taken to Mantova, where we were accommodated with numbers of other re-captured prisoners-of-war in some extremely spooky army transport garages on the edge of one of the freezing, fogbound lakes that surrounded the city. It was staffed by renegade South Africans who were either genuine sergeants or else had assumed the rank. They had invented for themselves a horrible pastiche of their real uniforms to demonstrate their sympathy with their new German masters, British khaki battledress, dyed black.

The temptation to do away with these monsters was very strong and we might have done so if the guards had been Italians, but they were not even Germans. They were Mongols, apostates from the Russian Army, dressed in German uniforms, hideously cruel descendants of Genghis Khan's wild horsemen who, in Italy, had already established a similar reputation to that enjoyed by the Goums, the Moroccans in the Free French Army, so we swallowed our pride but not our anger and contented ourselves with refusing to acknowledge their existence when they tried to give us orders.

Then one morning, at first light, we were hurried through icy streets to the railway station by tough German soldiers with large metal plaques on their chests announcing that they were *feldgendarmen*, military policemen, armed with Schmeisser machine-pistols, while those of the civilian population who were abroad at this hour hastened on their way with eyes averted. Later that day we crossed the Brenner and the Reich swallowed us up.

We were taken to an enormous prison camp, Stalag VIIA, at a place called Moosburg in the marsh lands north of Munich and about twenty-five miles from Dachau which was situated in the same sort of nightmare country. Over the entrance was a crude depiction of prisoners tottering into captivity with the words 'To Berlin' under it, which at the time seemed a typically heavy Teutonic joke. In it there were some eighteen thousand prisoners and it was said that more than twenty different nationalities were to be found in the low, grey huts which formed the living accommodation. In fact, Stalag VIIA was a microcosm of the world, or perhaps it would be more accurate to say a microcosm of hell. In it I saw the SS at work on Yugoslav peasants, men and women, many of them Muslims from Bosnia, stripping them naked and kicking them around in the snow. Later, when their day's work was done, I heard these same SS singing harmoniously together in their warm quarters, full of *gemütlichkeit*. I saw starving Russians who en masse resembled packs of wolves, each one with a minute sack on his back which contained his meagre possessions and which he never took off for fear of someone stealing them; men with otherwise nothing to lose who, when they laid hands on a guard dog which the Germans had been rash enough to let into their compound, skinned it, cooked it and hung the skin on the wire; men to whom we used to give bread whenever we could, for which they fought savagely among themselves.

It was here, at Moosburg, that I attended a theatrical

entertainment in which the equivalent of the stalls were packed with French officers in elegant uniforms and women in evening dress with elaborate hair-dos, who were not women at all but had rubber tits. At Moosburg the French, who were the oldest inhabitants, virtually controlled the inner workings of the camp. It was at Moosburg, too, that I met a band of soldiers, some of them British, who were subject to less stringent control than we were, being members of working parties, who used to go out through the wire at night dressed in black to render themselves less visible, to pleasure German women whose husbands were at the front or already dead, payment being made either in food or cash, or both. It was here, at Moosburg, that I really realized for the first time the extent of the degradation that the Nazis were spreading throughout Europe, and began to hate them for it.

Then one day, when time had ceased to have much meaning, some of us were dressed in strange Yugoslav uniforms and wooden clogs, herded into a train, and taken on an interminable rattling journey to what had been Czechoslovakia by way of Landshut, Regensburg, Hof, Plauen, Chemnitz, Dresden, Waldenburg, Glatz (now Klodsko in Poland), then by a network of minor lines to Trebovice and eventually to a place in what was then Silesia called by the Germans, who claimed it as part of the Reich, Märisch Trübau, by the Czechs Moravská Trebová.

So bad were the weather conditions in the Bohemian Forest through which we passed in the course of this journey that when we were allowed down on the line in order to relieve ourselves the guards did not even bother to guard us and I remember shedding tears from sheer frustration, standing alone in the blizzard trying to pluck up courage to run away but knowing that with wooden clogs on my feet, a Yugoslav uniform on my back and without food, money or maps, there was no point in doing so. I was safe in the heart of Fortress Europe.

The camp at Moravská Trebová, Oflag VIIIF, was housed in the former Czech Military Academy. Here we had Canadian Red Cross food parcels every week, the best of all parcels apart from those sent by the Scottish Red Cross. Here, the arts flourished and the theatre played to packed houses as it had done at Chieti, whose occupants had been transported en masse to Germany, thanks to their ridiculous senior British officer who had forbidden them to break out of the camp under pain of court-martial.

Dramatizations of prison camp life often give the impression that the aim of prisoners, one and all, was to escape, and that anyone who failed to try was in some ways lacking in moral fibre. This is nonsense. Only relatively few prisoners-of-war had the skills necessary to make a successful escape through enemy territory, and any Escape Committee that was not entirely irresponsible only allowed escape attempts by people who could prove that they possessed these skills. On the other hand, hundreds of prisoners who knew that they had no chance of making a successful escape, worked away on tunnels which they themselves would never use, dedicating a certain amount of time each day to digging, acting as look-outs and so on in the same way as they would devote a certain amount of time to reading, playing poker or attending lectures an accountancy or book-keeping.

Here also there were numbers of desperate, ruthless men, many of whom had been brought here without having savoured freedom from the almost escape-proof Italian punishment camp, PG5, which was situated in a hilltop castle at Gavi in the mountains between Genoa and Alessandria, and which German parachutists had surrounded on the day of the Italian armistice. Among them were David Stirling, captured in North Africa, and George Duncan who had led our raid in Sicily and had later walked out through the front gate at Chieti disguised as an Italian *soldato*.

Here, at Moravská Trebová, five separate tunnels were being dug simultaneously and hundreds of prisoners were involved in working on the various faces, operating air and water pumps and the underground tramways, which took the spoil from the face to the foot of the shafts from which it was hauled to the surface. The amount of earth extracted from these workings was prodigious and before the snow melted its disposal was a great problem. In fact the only place where it could be secreted was in the attics of an unused barrack block, which eventually collapsed under the weight, leaving us with the largest collective fine that we had so far suffered.

Later, when the snow melted, the earth was disposed of by prisoners who walked round the camp with sacks suspended round their necks beneath their coats, gently leaking it on to the ground as they pursued their endless, apparently aimless, perambulations. Other stooges, as they were known, acted as look-outs for the working parties and for The Canary, the clandestine radio that had been built in the camp and which defied every effort by the Germans to discover it even when, later that year, it moved with us to another camp.

In order to discover the locations of these tunnels and The Canary, the whereabouts of which was known only to a handful of prisoners, the Germans sent in what were called Ferrets. Gestapo and other lesser persons, disguised as workmen, and on one or two occasions these Ferrets were lured into cellars where keys were turned on them. When Ferrets entered the camp there was often no time to get the men working in the tunnels to the surface before the lids were replaced, and it was an eerie experience lying at the bottom of a shaft listening to the Ferrets as they hammered on the floor above, and often on the lid itself, while searching for it, and to hear their dogs panting and scratching.

Those lids were works of art. Some of them were the work of South African mining engineers – only a very few of the large numbers of South Africans in prisoner-of-war camps changed sides, in spite of all the inducements offered them by the Italians and Germans. To make a lid, a section of tiled floor, or whatever materials it was constructed with, was taken up and set in a block of concrete in such a way that if it was struck with a hammer it would give off the same sound as the surrounding floor. Special tools had to be devised to lift and replace these lids as they were extremely heavy.

As the weeks and months passed, one by one these tunnels were discovered, as they moved out towards and beyond the perimeter wire. In one instance a guard stamping up and down on his beat outside the wire fell through the roof of one and disappeared from view completely, which was highly diverting to those who saw it happen; but with these discoveries our morale fell too. That is, with the exception of the members of the Escape Committee – there was an escape committee in every prison camp. Although we who toiled away at their behest did not know it, these tunnels were intended to distract the enemy from a grander, more secret design, one far more serious in intent than a simple escape attempt with the purpose of enabling prisoners to rejoin their loved ones and their units.

What the committee planned was a mass escape of saboteurs and other men trained in irregular warfare in which the camp abounded who, when the escape actually took place, would be taken into hiding by the Czech underground movement whom they would then train in their various skills until such time as they could all take the field together against the Germans.

To further this scheme an Engineering Society was formed and the Germans were then asked to give permission for the making of large scale models to which they, rather surprisingly, agreed. One of these models was of a Bailey Bridge – a temporary bridge made of pre-fabricated steel parts that could be rapidly assembled and was widely used for military purposes. This one was built of wood and was of a very large scale indeed.

The intention, when the time for the break-out arrived, was to fuse the perimeter lights and searchlights and then drop the bridge on the wire. When this was done the entire band of would-be escapees would charge across it in the darkness into the wilds of Czechoslovakia. This desperate plan, which had much in common with the Charge of the Light Brigade and which cast serious doubts on the sanity of whoever conceived it, was in direct contravention of the Geneva Convention in relation to prisoners-of-war, which expressly forbade anyone who escaped from taking up arms. Fortunately for all concerned, including myself, it was never carried out. It was betrayed by an officer in the camp who was at least partly acquainted with it and who had the misfortune to have a wife and children living in occupied Europe and the Gestapo knew it. Not unnaturally he told them what he knew. Few of us, fortunately, have ever found ourselves in such a position, and it is difficult to say what any of us would have done in similar circumstances, faced with betraying an ostensibly crazy plan with very little chance of success or of sacrificing his wife and children.

The results were tragic. At this time two Czech-speaking British officers were got out of the camp with the intention of making advanced contact with the Czech underground. It may be that the Germans allowed their escape to happen. Nothing more was heard of these emissaries until one morning the Senior British Officer was told to present himself at the main gate where he was handed two tins containing their ashes.

Within a day or two of this tragic dénouement Oflag VIIIF was evacuated. We were marched through a village, in which all the inhabitants had been ordered to stay indoors with blinds or curtains drawn, by guards armed with submachine-guns, to a railway embankment where a large number of freight cars stood waiting on what was a spur of the line to Trebovice. Here we were lined up on the edge of the embankment and handcuffed. In view of what had happened to our two Czech-speaking officers the situation looked grim.

'You can't do this to us,' one veteran prisoner who had been captured in Norway in April 1940 shouted. 'We're British!'

He had been handcuffed years previously in a fortress at Thorn in Poland. Nevertheless his protest seemed curiously comic at this particular moment, as the majority of us were convinced that we were going to be mown down.

But we were not mown down. Instead, we were ordered into the freight cars, the doors were locked and we began another endless journey, protracted by air raids in the later stages, this time north-westwards to Brunswick by way of Görlitz, Dresden again, Leipzig, Halle and Magdeburg. At least we were travelling in the direction of home.

The floors of the freight cars were covered with straw and each one was provided with a barrel for us to pee in. We got the handcuffs off in about ten minutes and consigned them to the barrel. When it grew dark and the barrel was sufficiently full we asked the guards for permission to empty it down a steep embankment at one of the innumerable and interminable halts. Soon there were no handcuffed prisoners on the train at all.

16

Götterdämmerung
(1944–5)

Oflag 79, our new home, was at Braunschweig-Querum on the north-western outskirts of Brunswick. There we were accommodated in what had been the barracks of the Luftwaffe until the Allied bombers had made the place too hot for them to continue to live in it. One of the Luftwaffe's airfields was at a place called Waggum, about five hundred yards away on the other side of the *autobahn* from Berlin to Hanover and points west. So recently had this evacuation taken place that in one cellar signs of what must have been a wild farewell party were still in evidence: innumerable empty bottles, cigarette ends and cigar butts (which we appropriated), items of female underwear and others of an even more intimate nature. At Waggum, too, there was a military factory, soon to become a favourite target for American Fortresses, the Mühlenbau und Industrie Aktiengesellschaft which made aircraft parts and assembled ME110s. On the other side of the camp, a few hundred yards away, was the Neemo Aktiengesellschaft factory, in which hundreds of Russian and other slaves were engaged twenty-four hours a day in constructing parts for V1 and V2 weapons. Not far off, too, was the Bussing truck factory, a tank factory and an assault gun works. Surrounded by all these targets, which had been pinpointed by British Bomber Command and the Eighth American Airforce, it was difficult to believe that this situation had not been chosen for us by the Germans without some forethought. There was no view to speak of as there had been at Moravská Trebová in Czechoslovakia, of wide open Central European spaces, spaces into which every night, as the Russians advanced westwards, the whole camp, once in bed, would shout as one man, 'Come on, Joe!', to the fury of our hosts. Here we were hemmed in by nasty, dripping, Nordic woods.

In the first six months, during which we were in residence,

Allied aircraft were overhead on a hundred and sixty-nine days and nights out of a hundred and seventy-five; armadas of Fortresses of the Eighth American Airforce and attendant fighters, anything up to two hundred-strong by day, and by night RAF Lancasters and Halifaxes and Mosquitoes. The Mosquitoes were particularly unpleasant as they often came alone or in small numbers, and buzzed around for ages carrying one four-thousand-pound bomb with which they appeared to be trying to hit the Neemo factory, but without success. Meanwhile, we cowered in the flimsy cellars beneath our living quarters with our patriotism at a low ebb, listening to these projectiles as they came howling downwards, longing for the opportunity to dig ourselves slit trenches somewhere out beyond the wire which enclosed us. Altogether, up to 16 October, we were at the epicentre of thirteen heavy raids, not counting raids by Mosquitoes and four large fighter strafes. Of the heavy raids that of 24 August was the most memorable.

That morning the 102nd red air-raid warning sounded, which signified that the enemy – in this case our own side – were overhead. Almost immediately we saw the first wave of Fortresses, the harbingers of more than thirteen hundred aircraft, coming in at between twenty thousand and thirty thousand feet through colossal flak, beautiful as flying fish, shimmering in the sun, this time not bound as they more often were for points east but, as the markers which the leaders let fall almost lazily to earth indicated, for Waggum and Querum and, among other targets, for us.

Later that afternoon, an American crew member of one of these Fortresses, which had been shot down over Brunswick, was brought to the camp to begin his life as a prisoner-of-war, after he had escaped, landing in a tall tree, being lynched by the civil population, who by this time were far more bloodthirsty than the majority of the military. There he stood for some time, surveying the smoking ruins of the barrack blocks in which the Germans were still burrowing for the dead, without saying anything. By a miracle only three prisoners were killed and forty injured. The Germans guards had far higher casualties. Then he lit a Camel, a cigarette none of us had seen for years. 'Well,' he said, after inhaling deeply a couple of times, 'I guess you can't make an omelette without breaking an egg.'

Extracts from a copy of a letter sent to Lieutenant Anthony Simpson, Royal Artillery, on the day after the events just referred to but, not surprisingly perhaps, never received by him:

Oflag 79

25 August 1944

Dear Tony,

Tell your Mum [she was in some secret job in the War Office] to send us some anti-aircraft guns. What happened yesterday – see your paper of today's date – is beyond a joke. And tell her lots of ammunition, fused for whatever height Fortresses fly at. We're not prepared to put up with this sort of thing even from our Allies from 'across the herring pond'. However, it is not only our American cousins, as some people like to call them, who are to blame. A very large fragment of an enormous bomb which the RAF delivered to us the other night when salvaged was found to have the name of a firm in Birmingham on it, of which the father of one of my room-mates is the chairman. 'Daddy' is not very popular at the moment.

PS. *Don't* show this letter to my parents, just *tell your Mum to Stop it*!

The winter of 1944 was gruesome in every way. All through March and early April 1945 we wondered if the SS would whisk us away east of the Elbe to be used as bargaining pawns, or simply to be obliterated as those Polish officers had been at Katyn. Plans were made to take the machine-gun towers by storm under cover of a barrage of stones thrown by massed stone throwers; but we were saved from this unspeakable necessity by the intervention of an extremely distinguished-looking member of the German Foreign Office, who turned up one day and stayed on, hoping to save his skin. Hanover fell to the US Ninth Army on 10 April after a spirited resistance by Hitler Youth and Flak Troops, who depressed the muzzles of their anti-aircraft guns in order to take them on. Then, after a night in which Allies and Germans had lobbed shells over the camp at one another, rather as if they were playing tennis, at nine in the morning on 12 April, a nice sunny day, a jeep with two Ninth Army GIs and a French worker named Pierre drove up to the main gate and took the German surrender from the camp commandant, Von Strehler.

'We're looking forward to knocking shit out of the Russkys now,' these two amiable fellows said, and departed in the direction of Berlin. 'Technically' we were liberated.

Later that morning, when it became obvious that we were not likely to be going home for some time yet, I went out through a hole that some other seeker after freedom had already cut in the wire, and walked through dark woods of splintered pine trees. There were innumerable craters – a small part of the six thousand eight hundred tons of bombs dropped on Brunswick which had obliterated six hundred and fifty acres of the city but failed to destroy the camp. Standing on a bridge over the Berlin autobahn at a point where it crossed the railway line I could see what was left of the airfield, with some burnt-out skeletons of aircraft standing on it.

I had expected to see long lines of Ninth Army transport moving eastwards along the autobahn towards Berlin, but as far as the eye could see in both directions there was not a vehicle or a human being. To the south-east of Brunswick, beyond the woods that hid our prison camp from view, columns of black smoke were rising high in the sky and from beyond the horizon to the north and east came apocalyptic rumblings; but here, where I stood looking westwards across the endless flat fields of the Saxon plain, there was no sound except the twittering of the skylarks high above. They reminded me of the skylarks near Stonehenge and above the Purbeck Downs in Dorset; but all I felt was very, very old. I was liberated but I had none of the sensations of freedom.

I continued westwards along the autobahn in the direction of Hanover, passing under an exit bridge. I then continued not on the autobahn, where I might become the target of some fighter pilot idly testing his guns, but in the fields alongside. Then I crossed the Weser-Elbe canal, and on that, too, there was not a barge or a human being in sight. It was rather depressing country, with here and there a group of farm buildings or a small village, villages that I knew from studying the escape maps in the camp mostly had names that ended in '—*rode*' or '—*büttel*'. These features, together with some telephone lines that provided a link with the outside world, and an occasional plantation of trees, were the only things that broke the monotony of this great inland sea of plain which was so extensive it was said that the King of England could stand on the battlements of Windsor Castle and, looking eastwards down the Thames Valley, know that from where he stood there was no higher land until one reached the Urals. I wondered if he was doing it now. I had always had a soft spot for him ever since, high in the Apennines, I heard his broadcast on Christmas Day 1943, when in laboured but sincere tones he had sent a message from Sandringham to, among others, those on Italian peaks. 'Wherever you may be your thoughts will be in distant places and your hearts with those you love,' he had said. And I, together with someone else on hearing him, had not been ashamed to shed a tear.

Under the trees on the edge of one of these plantations, two or three hundred yards from the road, I could see something that gave me the shivers, a Tiger tank, hull down, apparently dug in, with its crew lounging around it, waiting. Waiting for what, I wondered? To make a last stand, someone to surrender to, an issue of cigarettes?

I crossed to the far side of the autobahn so that they would not see me and pick me off, and kept going westwards. Then, seeing an isolated farmhouse a couple of fields away, on impulse I walked across to it. Everything I did that morning was unpremeditated.

I went into the farmyard with a stone in my hand, expecting to be met by a savage dog, but there was none. I had felt fear when I saw the Tiger but now I had no feeling at all that what I was doing might be foolish, that there might be German soldiers inside the building, hiding, overrun, left behind by the battle but still effective. I was numb. I did not even stop to think why I was going into the house at all.

The farmhouse was of indeterminate age, built of that slightly shiny sort of brick which never changes its colour or texture from the time that it is made until the house falls down. I opened the door and found myself in a large kitchen, with a big stove to one side, in which what was presumably the entire family were seated at a long table, eating what must have been the equivalent in Saxony of English elevenses. Four masks by Bosch or Breughel, mouths full of cake, with little stone eyes, regarded me, neither friendly nor unfriendly – masks in a stone wall. They were the first German civilians I had ever met.

'*Kann ich essen, bitte?*' I asked in as amiable a manner as I could manage under their stolid gaze – old father, still robust, old matriarch, head of the family, fat son of forty-odd, thin wife, the men in greasy-looking black suits, the women in greasy black dresses and aprons, the working gear of peasants anywhere. No reply. They had stopped chewing. They even seemed to have stopped breathing. No sound except a long case clock with the name of the maker and the place of origin which ended in '—*rode*' engraved on its dial, tick-tocking in a corner. '*Kann ich essen, bitte?*' I said again, more loudly. By this time I was not interested in their bloody cake. It was a game.

Nothing.

The cake was not particularly attractive-looking, but it was round and large and sticky, and I had not set eyes on anything like it since the middle of 1942 in Egypt. It stood on a wooden platter with a big knife beside it. I picked up the knife, cut myself a modest slice and with the words, '*Viele danken,*' spoken in as ironic a tone as I could manage in German, turned on my heel and walked out of the door, expecting to feel the knife between my shoulder blades.

Outside I waited with my back to the wall for perhaps thirty seconds, which seemed longer, expecting to hear a babble of voices, let loose by my departure, before continuing on my way. But there was nothing. Just the wind in the telephone wires that stretched away from the house across the plain and the twittering of the larks overhead.

In one of the villages ending in '—*rode*', a very small place which appeared to be deserted, I met some Americans. Only the fact that the spotless blinds were drawn in the neat houses showed that the inhabitants were there all right, hidden behind them as they had been that day we were marched through Moravská Trebová. There were half a dozen Americans with a couple of jeeps between them, armed to the teeth and carrying their weapons with the sort of negligence that spoke of long familiarity and use. In one of the jeeps a radio crackled. I told them about the Tiger and they said the place was still crawling with Krauts – Hitler Youth mostly – and that I should watch my step. They were incurious about my being a prisoner-of-war, just as I had been incurious about prisoners-of-war before I was captured. Being a prisoner was something they could not envisage. They gave me a whole pack of Camels, some tins of Spam and some weird rations contained in what looked like oversize toothpaste tubes, of a sort that I had never seen before, which the affable man who gave them to me described as 'a loadashit'. Their morale was very high. They spoke of a possible show-down with the Russians east of the Elbe with positive relish. Just before I left, the sergeant in charge of the party who had told me to watch my step beckoned me round a corner.

'We just hanged a guy,' he said, in a matter-of-fact way, just as he might have said, 'We just had breakfast', pointing to something that looked like a sack of potatoes suspended by a piece of rope from a lamp-post and gently turning, first one way, then the other. 'Here we were, as friendly as might be, and he takes a shot at us, not one shot but a whole lotta shots. I tell ya, Lootenant, around here ya gotta watch ya step.' Suddenly I was glad I was wearing an Allied uniform.

I had had enough liberation for one day. I had never thought of the camp as being home before. Now I beat it back there, up the autobahn, over the canal, under the exit bridge to Brunswick West, back through the broken woods. I was liberated but I was still not free.

Telegram April 26, 10.30 a.m. Official.

Newby. 3 Castelnau Mansions, Barnes SW13.
Am in corrugated iron hut in wood somewhere in Sussex being given leave and clothing coupons. See you this afternoon, Love Eric.

After this I went home to Three Ther Mansions.

17

Commercial Traveller
(1946–54)

In the winter of 1945 I went back to Italy, borrowed a jeep, drove to Fontanellato through what were now the ruins of Italy and, in what had once been the stables of the ancient moated castle in which the last scion of the Sanvitale family still lived, asked Wanda to marry me. In spite of the fact that both of us were working for a branch of MI9, helping Italians who had helped prisoners-of-war to escape, the Allied authorities made things as difficult as possible for us to marry – on the grounds that my wife-to-be was an enemy alien, although in fact she was a Yugoslav, a Slovene from the Carso, whose father was a noted anti-Fascist.

We finally got married in the spring of 1946 in Florence, in Santa Croce, in the beautiful Bardi Chapel which is decorated by Giotto. Then, back in England, I started work in the family business of Lane and Newby Ltd, Wholesale Costumiers and Mantle Manufacturers.* I spent much of the next seven years acting as a commercial traveller on behalf of the firm, tottering up the backstairs of stores in London and the provinces with armfuls of stock to show to buyers who had gone on holiday, to coffee, to Paris, or the ladies' powder-room, had just been sacked and not yet been replaced, had gone mad or had something else wrong with them, had over-bought, had not yet started buying or simply didn't want anything of the kind I had to offer. Wherever I went I travelled with enormous wicker baskets containing coats and suits and with trunks containing dresses which I unpacked and re-packed at least twice a day, standing in a sea of tissue paper. Once, at Liverpool, I saw the whole collection come off the hook of the crane that was lifting it on

---

* Most of the material in this chapter has been taken from *Something Wholesale*, now unfortunately out of print, and likely to remain so, which is why I have ventured to reproduce it here.

board the Irish boat and fall into the Mersey. Fortunately it was a duplicate collection, adequately insured.

The Journey – the great journey as opposed to what might be described as lightning raids, a day visit to Southampton for example – took place twice a year. If it was made a week too early the buyers had not yet received their buying allowance. If it was a week too late they had probably spent it all. It lasted ten days or a fortnight and was carried out entirely by means of the British Railway system which, although already groggy, had not reached the depths of demoralization it was later to plumb.

There was a precise ritual connected with The Journey, in which future generations of savants who have the inclination to study it may find as much significance as did the author of The Golden Bough in the Slaying of the Priests at Nemi. It always began in the north and it was impossible to do good business in both Glasgow and Edinburgh. If the orders were good in Glasgow the buyers in Edinburgh were informed by some sort of bush telegraph and asserted the age-old hostility which exists between the two cities by buying as little as possible, and vice versa.

From the stock-rooms in the railway hotels in Glasgow and Edinburgh in those years before Beeching destroyed the railway system, it was possible to make daylight excursions by train to Dundee, Stirling, Perth, Ayr, Peebles, Berwick-on-Tweed and Aberdeen, returning to them the same night. Mr Wilkins, our senior traveller, claimed to have reached Inverness in one day, transacted business, and then returned to Glasgow, but no one really believed him! The further from his base the more the traveller's spirit failed him, like his medieval counterpart passing uneasily through country which on his map was marked 'Here be dragons'.

South of the border The Journey took in Newcastle, Manchester, Liverpool and the great industrial heart of England in which one of the most terrifying of all the buyers I ever encountered, Miss Trumpet of Throttle and Fumble in Yorkshire, a woman more than six feet high who dyed her hair bright orange, stood guard over her department wreathed in the smoke and flames of a thousand Bessemer converters like a ravening Fury at the mouth of Hell.*

This was The Journey. It petered out in the flat clay of the Midlands with a visit to Nottingham and Leicester. Nothing else in Britain was worth the expenditure of so much time, money and effort. Buyers in the Eastern Counties, made gelid by the winds that droned over them from the Urals, took every opportunity of deserting their patrimony and doing their buying in London. Most of the South Coast was so close to London that buyers commuted to it. Travelling into the eye of the setting sun towards the Cassiterides and Atlantis, phantom isles of the West, the softer the air grew, the more woolly-minded were the inhabitants. Decisions that were a matter of a moment in the North, in the West, in Bath or Exeter or Torquay or Plymouth were drawn out intolerably. It was like trying to swim through a sea of cotton-wool. Even Bristol, once famous for its swashbuckling merchants, had succumbed to the deadly softness of the Atlantic air. Beyond the Tamar, in Cornwall, apart from cream teas, the production of hideous pottery and equally awful objects embellished with poker-work, commerce was non-existent. Wales was another problem, as it always has been throughout history.

My first steps on The Journey were made in company with Mr Wilkins in 1946. After an appalling journey from King's Cross to Edinburgh, in which we sat upright throughout the night in a third-class compartment with five other occupants, two of whom were drunk, we were decanted on to the platform at the Waverley Station from which porters of the North British Hotel hurried us with a great trolley-load of our skips and trunks to a stock-room in the lower parts of the building, a tall narrow room illuminated by a fifty-watt bulb. Besides a number of rails on which our 'models' were to be hung, the furniture consisted of a number of rather rickety cane chairs and two trestle tables covered with white sheets that had been neatly patched. The effect was of a mortuary or a place where members of the Reformed Church might pray together before proceeding to England by train. The view from the window embraced the roof of the Waverley Station and a single span of the North Bridge.

* The names of buyers and of the firms they worked for have been changed for obvious reasons.

At intervals the entire prospect was blotted out by clouds of smoke emitted by steam locomotives revving up on the rails below. It was half past eight. Our first customer was at nine-thirty. We had not yet shaved, and had had no food since lunchtime the previous day.

Breakfast resembled a slow motion film of a coronation. At intervals rather niffy waitresses brought food, but with none of the supporting ingredients that would have transformed it into a breakfast – porridge came without milk, margarine and marmalade without toast to spread them on, tea without teacups and, presumably in obedience to some not yet defunct war-time regulation, there was only one bowl of sugar to four tables. By the time we had finished breakfast, having half risen in our seats despairingly half a dozen times in attempts to attract the attention of these ladies, it was ten past nine.

There was no time to wash or shave. We raced to the stock-room. I was in charge of the dresses, Mr Wilkins had the more robust coats and suits. To me it seemed inconceivable what havoc a night in trunks had wrought among my fragile garments, in spite of their having been carefully interleaved with tissue paper.

By the time we had finished hanging the stuff up it was nine twenty-seven. Mr Wilkins handed me a sheet of writing-paper, part of a large supply he had already filched from one of the hotel writing-rooms, on which he had drawn up a time-table:

9.30	Mrs McHaggart, Robertson's, Edinburgh
10.30	Mrs McHavers, Lookies, Dundee
11.00	Miss McTush, Campbells, Edinburgh
11.45	Mrs McRobbie, Alexander McGregor, Edinburgh
2.30	Miss Wilkie, McNoons of Perth
4.30	Miss Reekie, Madame Vera, Edinburgh

To me it sounded more like a gathering of clans in some rainswept glen rather than a series of assignations to buy dresses in the sub-basement of a railway hotel.

In the three minutes that remained before the arrival of Mrs McHaggart, Mr Wilkins treated me to a brief, brilliant summing-up of their idiosyncracies, which my parents had already described to me at some length; but by the time he had finished I was in such a state of apprehension that I could scarcely distinguish a McHaggart from a McTush.

'Mrs McHaggart is a good buyer but she doesn't like us to serve any of the other stores in Princes Street. Of course we do – it wouldn't be worth coming here if we didn't – and she knows we do. The buyers here know everything. They all have relatives in one another's shops.

'What we have to do is to get Mrs McHaggart's order down on paper. If it's good enough we don't show the styles she's ordered to Miss McTush. They're enemies. If we get a poor order from Mrs McHaggart then we show everything to Miss McTush and change the styles. Miss McTush knows we do this so we can't change them very much. Mrs McRobbie is the same as Mrs McTaggart and Miss McTush. She's in Princes Street, too. The most important thing is to keep them from meeting. If they do at least one of them won't give us an order. That's why I've put in Mrs McHavers between Mrs McHaggart and Miss McTush, because she comes from Dundee, but Mrs McHaggart is up to every kind of dodge. She often leaves her umbrella behind after she's put down her order to give her the excuse of coming back and finding out who else we're serving. Miss McTush doesn't really mind what Mrs McHaggart and Mrs McRobbie buy as long as she gets her delivery before they do. In fact we deliver them all at the same time – we wouldn't dare do otherwise – so Miss McTush is just as difficult as the others. Mrs McHaggart only buys Coats and Suits and Two-pieces. She's not supposed to buy Two-pieces but she does. That's why you won't see Miss Cameron, the Dress Buyer. Miss McTush buys everything. Mrs McRobbie buys everything. Miss Reekie can buy anything but usually she buys nothing. She's a most difficult woman. I call her 'The Old Stinker' on account of her name being Reekie,' said Mr Wilkins. 'I usually take Miss McTush and Mrs McHavers out to lunch together because Mrs McHavers comes from Dundee and Miss McTush doesn't mind that. On Tuesday I take Mrs McHaggart. First thing on Tuesday morning I call on the ones who haven't given us an appointment and do some telephoning. With luck we see some of them in the afternoon or on Wednesday morning and I usually manage to get off to Glasgow on Wednesday afternoon for an appointment after the shops close in the evening. Some time today we have to do some telephoning to Galashiels and Hawick.'

'Don't you give Mrs McRobbie lunch?'

'Mrs McRobbie's got an ulcer. She never eats lunch. I like Mrs McRobbie,' said Mr Wilkins.

'What about the evenings?'

'If you want to take Buyers out in the evening, Mr Eric, that's your affair,' said Mr Wilkins. 'Personally I drink beer.'

As he said this there was a murmuring sound outside the door and Mrs McHaggart appeared.

Mrs McHaggart was tall and thin. She was invested with an air of preternatural gloom, accentuated by a small drip on the end of her nose. In all the years I was to have dealings with Mrs McTaggart the drop never actually dripped but always remained suspended on the point of doing so. She was dressed in claret-coloured tweed and jacket of a fur that was unknown to me, possibly made from the skins of animals trapped north of the Highland Line and over which hung an aura compounded of moth balls and Parma Violets. She asked after my parents in a kindly way, but her manner of doing so suggested that they were either dying or already dead and no one had informed her of the fact.

We were off.

18

Travels in My Imagination
(1947)

I wanted to travel to far distant places but the nature of my job made it unlikely that I would do so. Above all, I longed to visit Istanbul.

Sometimes the ideas I have formed about a place that I have never visited but long to visit are produced not by looking at photographs, as I had done when at a tender age I acquired *The Children's Colour Book of Lands and Peoples*, but by studying works of art. Once this has happened, however much the place has ceased to resemble the place which the artist may have imagined it to be when he depicted it, if it ever did resemble it, then this vision of it is indelibly fixed in my mind and I can never obliterate it.

In Istanbul this evocation of a Constantinople that possibly never was – its name was only changed to Istanbul in 1930 – a place I never visited until 1956, was performed for me back in 1947 by two nineteenth-century artists. William Henry Bartlett,* born in 1809 in Kentish Town, and Thomas Allom, born in London in 1804. I bought the books which were

*The engravings made from Bartlett's drawings of Constantinople first appeared in *The Beauties of the Bosphorus* by Miss Julia Pardoe, published in 1839. Miss Pardoe was the daughter of a retired major who had fought in the Peninsula and at Waterloo. She was 'a fairy-footed, fair-haired, laughing girl' whose first work, a book of poems, was published when she was only fourteen. She first went abroad to avoid consumption and in 1835 accompanied her father to Constantinople. Two years later, at the age of thirty-two, she wrote a best seller, *City of the Sultan*, which was swiftly followed by *The Romance of the Harem* and *The Beauties of the Bosphorus*. The engravings made from Allom's drawings of Constantinople were published in *Constantinople and the Scenery of the Seven Churches of Asia Minor* by the Rev. Robert Walsh, also published in 1839. Walsh, an Irish curate, was already over forty when he produced his *History of the City of Dublin* which appeared in two massive quarto volumes in 1815. Five years later he became chaplain to the British Embassy at Constantinople and this gave him the opportunity to travel in Turkey and other parts of Asia.

illustrated by them from barrows in London street markets. Both these artists were articled to architects, which accounts for their masterly treatment of buildings; both, like Edward Lear, their contemporary, were indefatigable travellers, often in wild and dangerous places.

The city of Constantinople and its environs, as depicted by Bartlett and Allom, were unnatural, in the sense that they were invested with an extraordinary silence and stillness which is foreign to the Orient, except in the hour before the dawn or when the sun is at its zenith, or in the desert. The inhabitants, men, women and children, were all frozen in the moment in which they were observed or imagined.

As in *The Children's Colour Book of Lands and Peoples* it was a world in which all was well. There was no violence. No heads of pashas displayed on dishes stood on the middle pillar, reserved for this purpose in the first court of the Grand Seraglio, beyond the Bab-i-Humayun, the Imperial Gate; no heads or other portions of lesser persons, such as ears and noses, were on display in the niches on its outward walls, as was customary at that time. And in fact here in this first court, apart from an occasional scream when the Executioner, who was also the Head Gardener, carried out a decapitation, there was always silence, none of the hullabaloo that arose in the second courtyard when the Janissaries, the always dissatisfied infantry of the Sultan's Imperial Court, reversed their great copper cooking bottles and beat on them to announce some fresh grievance.

'Anybody may enter the first Court of the Seraglio,' wrote Joseph Pitton de Tournefort, the famous botanist who visited Constantinople in about 1700 during a three-year journey through Greece and Asia Minor, in the course of which he made an extensive collection of plants for Louis XIV '. . . but everything is so still, the Motion of a Fly might be heard in a manner; and if anyone should presume to raise his voice ever so little, or shew the least want of Respect to the Mansion-place of their Emperor he would instantly have the Bastinado by the Officers that go the rounds; nay the very Horses seem to know where they are, and no doubt they are taught to tread softer than in the streets.'

In fact, the engravings of Allom and Bartlett's works gave no inkling of the violence and cruelty practised in the Topkapi Serai, until its abandonment between 1850 and 1853, in the reign of the drunken and wildly extravagant Sultan Abdul Mejid, who was so enfeebled by excessive indulgence in the pleasures offered by his Harem that he was unable to enjoy even the incomparable views from his palace across the Bosphorus.

They did not show, for example, odalisques being drowned in weighted sacks off Seraglio Point. The mad Ibrahim, the last Sultan with the blood of the Caliph Osman in his veins, who consumed amber dissolved in coffee and drenched his beard with ambergris, had two hundred and eighty of his women disposed of in this way with as little compunction as a motorist ordering an oil change. Nor did they show, and a good thing too, the newly-sheared African eunuchs being bastinadoed with drumsticks by their seniors – presumably like prefects dishing out punishment at an English public school to new boys – in the creepy, three-storeyed quarters of the Black Eunuchs, which are more like a deep ditch with a roof on it than a human habitation. Perhaps it was the only way they could warm themselves in what, for five months of the year, is a perishingly cold place.*

Nor did they show the hideous Seraglio deaf-mutes with their slit tongues and punctured ear-drums, whose principal *raison d'être* was to act as stranglers using the bow-string or, if they were strangling a Crown Prince in the Kafes, a silken one.

The Kafes, the Cage, was the suite of rooms in which the Crown Prince was kept captive during the life of the reigning Sultan. Ibrahim was in the Kafes from the age of two until he became Sultan at twenty-four. No wonder he was as mad as a hatter. Eventually he returned to it to be strangled. Osman III spent fifty years in the Kafes.

* 'The great manufactory of eunuchs visited by the explorer John Ludwig Burckhardt in 1813 was at Zawyet ed-deyr on the Nile near Assiut, 'which,' Burckhardt wrote, 'supplied all European and the greater part of Asiatic Turkey with these guardians of female virtue.' There, two Coptic Christian monks castrated about a hundred and fifty young Negro boys a year, post-operative treatment being to bury them up to their haunches in warm manure. Of sixty who were operated on in the autumn that Burckhardt was at Zawyet ed-deyr two died, which was regarded as an abnormally high mortality rate.

In fact Bartlett and Allom recorded none of the scenes on which a present day photographer would turn his cameras with relish. There was not even a single burning building to be seen in their pictures, which is strange because the history of Constantinople from the point of view of underwriters and firemen has been one of unending employment. These conflagrations were all due to the Turkish predilection for smoking in bed, forgetting to snuff candles, carrying hot coals with wooden pincers from one room to another in their wooden houses, drying linen over braziers, engaging in mass fry-ups with oil during what is known as the Egg-Plant Summer (the equivalent of our Indian Summer) and simple arson.

With Bartlett and Allom one was in the springtime of the world, or else in a golden autumn. Whichever season, they were days of calm. In the Golden Horn there were ships at anchor with cock-billed yards, and from the shore little rowing-boats with prows as fine as needles moved out towards them without leaving a ripple on water which was as smooth as glass, while out in the centre of the stream one of the Sultan's thirty-two oared galleys was being rowed across to Galata from the Summer Harem down by Seraglio Point. Behind it, high in the background, rose the incomparable domes and minarets of the city, while along the banks of the Bosphorus on the European and Asiatic shores semi-comatose figures reclined on the balconies of the *Yalis*, the wooden houses built out on piles over the water. They were in the state of euphoria peculiar to Muslim gentlemen and described minutely by the traveller and explorer, Richard Burton, known as *Kif*.

Across the water from under the Imperial Gate of the Seraglio, a troop of horsemen wearing tall tapered hats of pale felt, were riding out past a fountain ornamented with gilded arabesques and crowned with domes, while nearby a servant was bringing, on his head, a coffee table laden with cups to a group of men smoking water pipes under a tree. In the cemeteries of Eyüp and Scutari, the one in Europe the other, Karaca Ahmet, the greatest necropolis in all Asia, women enveloped in the *feridge* and the *yashmak* crouched by the tall, slender headstones which mark the tombs, while in the distance a funeral procession wound away among the dark stands of cypress. These headstones were surmounted with carved stone headdresses, veils or hats for the women, turbans for the men and by the way in which these turbans were made up it was possible for those who had studied the matter to know whether the wearer was a pasha, a dervish, a eunuch or an ordinary, unremarkable man. Seeing them in their pallid, serried ranks under the funereal trees it was as if the dead had been resurrected, only to be turned to stone. Here, too, under a *baldacchino* or canopy among the cypresses, the favourite horse of Sultan Mahmud I was buried.

At Eyüp the new Sultan had come on horseback from the Seraglio to the Mosque there, the holiest in Constantinople, for the ceremony of girding the sword, the equivalent to his coronation, riding out from the Gate of the Shawl and the Curtain Gate. There, in the narrow street which ran out among the tombs to the great marble *türbe*, mausoleum, of Eba Eyüp (which housed the remains of a companion of the Prophet who is supposed to have died in the first Arab siege of Constantinople between 674 and 678), his subjects abased themselves.

At the Sweet Waters of Europe, on the banks of the Barbyses, at the head of the Golden Horn, unveiled Greek girls were dancing the *romaika* on the bank. It was St George's Day. In a harem a pale Circassian slave, from the shores of the Black Sea and attended by a black eunuch who was wild with jealousy, plucked languidly at a sort of lute while another, equally languid and sensual, sat at the feet of her master, who was smoking a *narghile*.

'There is a certain ferocity and irreclaimable wildness observable in a Circassian beauty. She gratifies the sensuality, but never secures the esteem, of him to whom she is afterwards consigned,' wrote the clerical author of *Constantinople and the Seven Churches of Asia Minor*. 'She is an object of desire, but never of regard, and always excites more fear than love . . . The splendour of the harem is contrasted with their own miserable huts (in Circassia); the rich stuffs in which they are clothed, with their homely, coarse, and squalid garments . . . they have no ties to attach them to their native land, or dim the bright prospect that awaits them in another. They look upon their sale to a foreign merchant to be the foundation of their future fortune, and their entrance into a foreign ship their first step to a life of pleasure and enjoyment; nor are they disappointed.'

These two Circassians, desirable as they were, were only two of many who had been rejected by the Imperial Chief Black Eunuch, the Kizlar Agasi, the Keeper of the Girls, as not having the necessary qualities for a Sultan's bedfellow. He, having done so, would have sent them to the Aurut Bazaar, or Female Slave Market, which, at that time (in the 1830s) was situated near the Burnt Column which marked the site of Constantine's Forum. As late as the 1900s it still functioned, but at Topkhaneh, beyond Galata, where Circassian merchants still carried on a brisk trade in their own flesh and blood, as well as Nubians and Abyssinians.

But what Allom showed was the original market, by the Burnt Column, a courtyard surrounded by balconies with lattices in which the white slaves of Georgia, Mingrelia (a mountainous area north-east of the Black Sea) and Greece awaited a purchaser. The price for a young white slave was then about 6000 piastres, £100, for a black slave about 200 piastres, £16. Yet,* as depicted by Allom and described by the almost lubricious Reverend Walsh, it was not a particularly melancholy scene, for the Negresses appeared as anxious as the Circassians to find a purchaser. Only the Greek girls formed an exception. 'Refined by education, strongly attached to their families, and abhorrent to slavery, their natural vivacity is overcome by their state, and they appear sad and dejected amid the levity that surrounds them.'

It was the moment when a sale was in progress. A bargain was about to be made; the Franks were privileged to be present for until recently a *firman* of the Sultan had excluded them from the Yessir Bazari, the Slave Markets, because it was thought that they sometimes bought in order to give the slave his or her freedom, something a still-young slave, at this stage of his or her career, particularly a black one, actively disliked. Now they were admitted; but only to satisfy their curiosity, not to traffic. On the left a veiled Turkish lady was examining the black female slaves, rather as if they were fatstock, which indeed they resembled. Certainly they were not the lissom figures of the popular imagination. On the right, the Circassians – there could have been no Greeks among them, as they were displaying their charms quite openly – were being offered by the slave-merchant (perhaps a Jew for he was bearded in an un-Turkish fashion and wore a fur kalpack on his head) to the master of a household, attended by his eunuch. I would never know the outcome; but this was the fascination of seeing the city in this way, for no transaction, no human action, was ever completed.

With these artists I could enter a *hamman*. 'It has been truly said of the Turks,' the author of *Constantinople* wrote, quoting another, unnamed authority, 'that "they hold impurity of the body in greater detestation than impurity of the mind." . . . They make frequent ablution so essential, that "without it prayer will be of no value in the eyes of God" . . . The law enumerates *eleven* occurrences after which the person must wash, some of which are exceedingly curious, but not fit for the public eye.'

What I was looking at was a male bath. To see the interior of a female one I would have had to go back in time to 1787 when d'Ohsson's *Tableau Général de L'Empire Othoman* was published, and looked at one of the engravings in that rather rare work. In it the women wore high pattens, sometimes covered in silver, sometimes jewelled, reminiscent of those worn by sixteenth-century Venetian prostitutes.

In the female bath there were almost as many children as women. It was an animated scene, but presumably of the artist's imagination, who could scarcely have been admitted. 'The mysteries of a female bath it is not permitted to see, no more than those of Eleusis . . . Their bath is the great coffee-house, where they assemble, and enjoy a freedom they can nowhere else indulge. If a stranger enter this sacred place by mistake, even his mistake is punished with death.'

Miss Pardoe's description of a bath to which she had the entrée was published some fifty years later. It was as if she was writing an extended caption to Bartlett's drawing of the same scene which he had, somehow, already witnessed.

'For the first few moments I was bewildered; the heavy, dense, sulphurous vapour that filled the place, and almost suffocated me – the wild shrill cries of the slaves pealing through the reverberating domes of the bathing-halls, enough to awaken the very marble with which they were lined – the subdued laughter and whispered conversations of their mistresses, murmuring along in an undercurrent of sound – the sight of nearly three hundred women, only partially dressed, and that in fine linen so perfectly saturated with vapour that it revealed the whole outline of the figure – the busy slaves passing and re-passing, naked from the waist upwards, and with their arms folded upon their bosoms, balancing on their heads piles of fringed or embroidered napkins – groups of lovely girls, laughing, chattering, and refreshing themselves with sweetmeats, sherbet and lemonade – parties of playful children, apparently quite indifferent to the dense atmosphere which made me struggle for breath – and, to crown all, the sudden bursting forth of a chorus of voices into one of the wildest and shrillest of Turkish melodies, that was caught up and flung back by the echoes of the vast hall . . .'

In the men's *hammam*, visited by Allom, there were fewer bathers and apart from the slapping of flesh, the cracking of joints as the *tellahs*, the bathing attendants, seized the bathers in an excrutiating lock, and the sounds of running water, the usual brooding silence reigned. Both the attendants and their victims had the foreparts of their heads shaven in the curious fashion of the time, leaving the hair long over the nape of the neck so that if it was only a little longer it could have been plaited into a pigtail. In fact, it was a very old hairstyle. It probably dated back to the thirteenth century when what was then a single, small tribe of Turks fled westwards before the Mongols of Genghis Khan, who themselves had a curious way with their hair, and the Turks may have adapted it to their own use. At his first meeting with Mongols, three days eastwards of Sudak in the Crimea in May 1253, the Flemish Franciscan, William de Rubruquis (Rubruck), on his way to visit Mangu Khan at Karakorum, noted that the Mongol men shaved a square patch on top of their heads, leaving a tuft to fall over their eyebrows, and longer hair at the back and sides.

The room was illuminated by long, parallel shafts of light admitted by the perforations in the dome which were filled with inverted hemispheres of glass and the steam and the wet floors produced a strange, unearthly radiance. This and the huge, shaven skulls of the *tellahs* gave the impression that what one was looking at was in another world.

This unearthliness which seemed to be a characteristic of Constantinople persisted up the Bosphorus, in the meadows of what are known as the Sweet Waters of Asia, the Göksu, the Sky Stream and the Kücüksu, the Little Stream, which lie outside the village of Anadolu Hisar, the Anatolian Castle on the Asiatic side. There the *arrhubas*, carriages with oval windows drawn by long-horned oxen, had already arrived in the meadows, laden with veiled women. It was Friday evening, the Muslim holiday, and these women were from the Asiatic side. The others, and the meadow was filled with them, would have come out of Europe across the Bosphorus in caïques. They sat around a fountain drinking sherbet, which in this case was made with the dried fruit of raisins, pears, peaches, prunes and so on, while the men sat apart.

One could have a better idea of what the *arrhuba* and its veiled occupants were like by returning once more to the Sweet Waters of Europe. There one could see the Asma Sultana driving through the meadows from her palace at Eyüp in her state *arrhuba*. It was a huge and heavy machine, carved with arabesques and springless; but in spite of its size, because of its oval windows which were so large that apart from the roof it was virtually an open carriage, the effect was one of extraordinary lightness, as if it and the carriages which followed with various other women of the seraglio, were some kind of exotic flying-machines which had just landed and were now being drawn away to Gate Number Seven where their occupants could disembark and go through immigration and customs. In fact it seemed as if it was only the horned oxen which chained the *arrhuba* to the earth.

The hair over the brows of these beasts was dyed with henna, and slender bow-shaped rods attached to their yokes were hung with tassels; sometimes the animals' tails were attached to the ends of these rods so that the tassels trembled even more than they would have done with the normal undulations of the road; and around their necks they wore bright blue beads as amulets to ward off the evil eye.

The Sultana's *arrhuba*, as were the others, was escorted by a grumpy-looking Greek *Arrhubagee* who had the same curiously shaven head as the men in the baths; and the curtains were drawn back, for the journey was nearly over and the Asma Sultana looked out with bold eyes above her white veil and a lock of hair peeped out from below her white coif. It was as if she was daring one to pay her a compliment; but behind her *arrhuba* was one of the black eunuchs who had a sword stuck in his cummerbund, and to do so would have meant instant death.

And I saw the bazaars, the greatest of which was the Kapali Carsisi, the Great Covered Bazaar, the oldest the Sahaflar Carsisi, the Market of Second-Hand Booksellers, which stood on the site of the Chartoprateia, the Book and Paper Market, which flourished when Istanbul was Byzantium.

Because in a bazaar one is normally more interested in the merchandise offered than in the surroundings, however impressive, as Bartlett and Allom showed them to be, one still wanted to know what the Great Covered Bazaar was really like in the twenty years before the Crimean War, before the Industrial Revolution swamped it with cheap trash and before the Grande Rue de Péra, now Istiklal Caddesi, which bisects the heights of Pera in the European City across the Golden Horn, siphoned off the fashionable shoppers who had previously patronized it in the seventies. To do so it was necessary to continue to consult the popular writings of Miss Julia Pardoe and the Rev. Robert Walsh. Both had, whenever the opportunity presented itself, which was not seldom in Constantinople, a slightly but nonetheless perceptibly risqué approach to whatever their chosen subjects happened to be.

Both saw the Great Bazaar when it was full of the perfumes of attar of roses, essence of lemon, extract of jasmine, as well as more aphrodisiac scents, such as those of the musky secretions of rats' tails and those given off by odiferous gums, which were much appreciated by the ladies of the harems who smoked them in pastille form, using pipes ornamented with turquoise and brilliants, which were anything up to twelve feet long, putting to their lips the mouthpieces of amber from the Baltic which were reputed to preserve the users from the plague.

In the Eski Bedesten, a fortress-like building in the heart of the Bazaar in which from time immemorial the most valuable merchandise had been offered for sale, and in which merchants wearing huge turbans sat writing with reed pens on yellow, parchment-like paper, they saw drinking cups lipped with precious stones, rosaries in which the beads were jewels, aigrettes of diamonds, and bridles covered with pearls, and gold rings. In the Kürkgüler Carsisi, the Furriers' Market, leopard skin saddle-cloths were on sale.

In the Sandel Bedesten, like its namesake the Eski Bedesten originally built of wood in the reign of Fateh Mahomet the Conqueror, not long after he took the city in 1453, they saw silks embroidered in thick gold and silver, the work of Armenian women who were even more strictly enclosed in their harems than in the Turkish ones. The silks from Bursa in Asia Minor were so heavy that it seemed impossible to Europeans that they were not mixed with cotton; and there were stuffs from Baghdad and muslins so thin that they were almost impalpable. Rebuilt in stone, as was the Eski Bedesten in 1701, the Sandel Bedesten was a huge, vaulted echoing building of Piranesian proportions and grandeur.

In the Street of the Shawl-Makers the most prized shawls came from Lahore and Tibet; the Street of Mirrors, the Aynacilar Sokak, sold little looking-glasses of gold and silver, or else they were velvet-covered and embellished with pearls from the nearby Street of the Pearl Merchants, the Incicler Sokak; the Street of the Slipper-Makers, the Terlikciler Soklak, was like 'a bed of tulips', Miss Pardoe wrote, full of gold-embroidered and jewel-sprinkled harem slippers and boots of bright yellow morocco for the Turks, crimson for Armenians, purple for Jews, black for Greeks.

In the darkness of the Armoury Bazaar the walls were lined with pieces of armour for men and horses, shields, helmets, spears, suits of chainmail, Indian bows, Tower muskets, American rifles, scimitars, Albanian pistols, Damascus sabres and so on.

It was the end of an epoch, and Allom and Bartlett, Miss Pardoe and the Rev. Walsh all knew it.

'Constantinople, having for centuries exhibited the singular and extraordinary spectacle of a Mahommedan town in a Christian region, and stood still while all about it were

Left: The traveller in his pram. Probably Broadstairs, 1920.

Below: My mother in our Napier car in Surrey, 1924.

Left: In the doorway of the cottage from which I set off on my first unaccompanied travels. Branscombe, 1925.

Below: In a charabanc somewhere on the shores of the English Channel, 1925.

1938–9: March 1939: storm in the Southern Ocean. 51° 12S. 158° 34W. Fifteen days west of Cape Horn, wind WSW Force 11. *Moshulu* carrying lower topsail and foresail; looking aft from the fore yardarm.

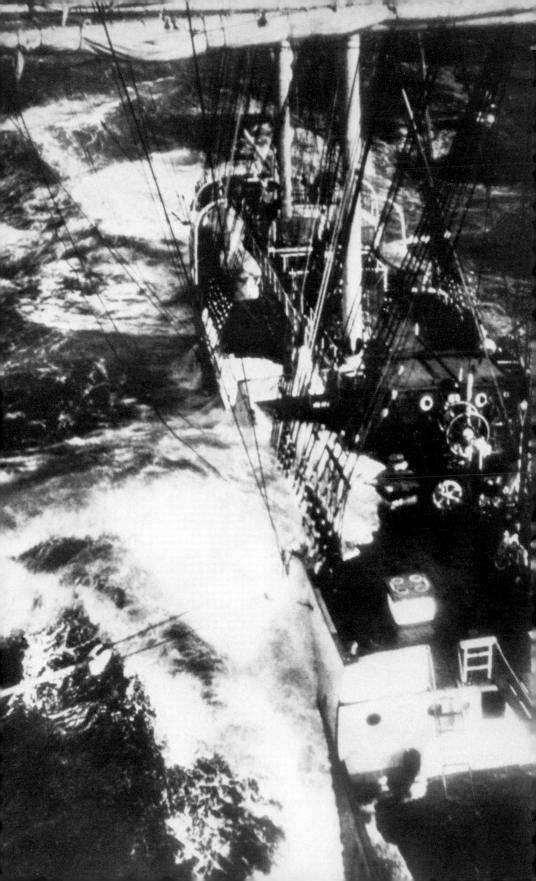

Washing up in the port fo'c'sle.

Wanda aged eighteen.
A graduation portrait,
1940.

On the run in Italy with
Donald Shaw and 'Doc'
Caraher. Winter, 1943.

Myself aged twenty-three.
Identity card photograph,
1942.

Constantinople in the 1830s: Seraglio Point, from a drawing by
W. H. Bartlett.

Portage on the Ganges, Winter, 1963. Wanda descending the river
with a supply of rice.

Above : Seen from the
wagon-lits window of the
Istanbul Express, 4.25 a.m.,
Svilengrad, Bulgaria.
Winter, 1964.

Right : Sheikh Ayid Awad
Zalabin in his encampment
in Wadi Rumm, 1967.

Right : Pera Palas Hotel,
Istanbul : the lift.

Below : Bath night at the
Pera Palas.

Riding from the Wash to Wimbledon by way of the Severn, without using classified roads. 1970.

Left: St Katharine's
monastery, Mount Sinai,
1971.

Below: The favourite wife of
Sheikh el Sheikh Abu
Abdullah of the Umzeini
Bedu. Sinai, 1971.

advancing in the march of improvement, has at length, as suddenly as unexpectedly, been roused from its slumbering stupidity,' the last wrote in 1839. 'The city and its inhabitants are daily undergoing a change as extraordinary as unhoped for; and the present generation will see with astonishment that revolution of usages and opinions, during a single life, which has not happened in any other country in revolving centuries. It is thus that the former state of things is hurrying away, and he who visits the capital to witness the singularities that marked it will be disappointed. It is true it possesses beauties which no revolution of opinions, or change of events, can alter. Its seven Romantic hills, its Golden Horn, its lovely Bosphorus, its exuberant vegetation, its robust and comely people, will still exist, as the tapered minaret, the shouting muezzin, the vast cemetery, the gigantic cypress, the snow-white turban, the *beniche* of vivid colours, the feature-covering *yashmak*, the light caïque, the clumsy *arrhubas*, the arched bazaar – all the distinctive peculiarities of a Turkish town – will soon merge into the uniformity of European things and, if the innovations proceed as rapidly as it has hitherto done, leave scarce a trace behind them.'

I had been born a hundred years too late.

19

When Did You Last Cross
the Oxus?
(1956)

In 1954 I left the family business and went to work for Worth Paquin, then a well-known couture house in Grosvenor Street.

While working away for Worth I wrote my first book, *The Last Grain Race*, based on the journal I had kept while serving as an apprentice in the *Moshulu*. By the autumn of 1955 it was finished and, to everyone's surprise, it began to show signs, long before its publication in September 1956, that it might be successful.

Some time in the spring of 1956 my publishers, Secker & Warburg, sent me a questionnaire to fill in for publicity purposes. The last question was, 'What would you most like to do?' and against it I wrote, 'Make a journey through unknown territory.' One day David Farrer, a director of the firm, asked me to go and see him. On his desk was the questionnaire. 'Just read this,' he said, handing it to me. Against my answer to the question 'What would you most like to do,' he had added in red ink, 'And so you shall!'

Meanwhile things were not going at all well at Worth Paquin, either for the couture house itself or for Newby. I had a gruesome interview with one of the directors, who said that they were keeping me on for the time being but were making no promises for the future. Emboldened by the reception my book was having, I told him that I had just had a book accepted for publication and that I was staying on for the time being and was making no promises for the future either. After this, during the lunch hour, I went into the Post Office in Mount Street and sent a cable to Hugh Carless, a friend of mine at the British Embassy, Rio de Janeiro, which read, 'Can you travel Nuristan June?'

Before being posted to South America Hugh, as a third secretary at the embassy in Kabul, had been able to indulge a passion for travels in the remote interior of Afghanistan. While engaged in one of these arduous journeys he had reconnoitred a twenty-thousand-foot mountain on the borders of Nuristan, otherwise the Country of Light, a region into which no Englishman had penetrated since Sir George Robertson, the British Political Agent in Gilgit, had explored parts of it in 1890–1, at which time it was known as Kafiristan, the Country of the Unbelievers. It was Hugh's ambition actually to get into Nuristan itself, and over the years he had written long letters to me about the country and its people.

It was during a rehearsal for the showing of the 1956 wholesale collection – which was attended by the suit buyer of a famous New York department store who, as it subsequently transpired, was not there to buy anything but to acquire 'as a pourboire' (as her agent described it) an oversize bottle of 'Je Reviens', the firm's scent – that I received a reply to my cable from Hugh in Rio: 'Of course, Hugh.' The story of our subsequent adventures is told in *A Short Walk in the Hindu Kush*.

I was now thirty-seven years old and my hair was beginning to disappear at an alarming rate. On my return to London from the Hindu Kush in the autumn of 1956 I decided to pay a visit to Mr Alexis (Alexis is not his real name) who ran, and still runs, a fashionable barber's shop in Jermyn Street; not in the hope that he would be able to arrest the process but that he might lay out what hair I still possessed in a way that would conceal the bald bits.

'Been abroad lately?' said Mr Alexis, as he began to work one of his miracles, affecting not to notice that I was trembling like a leaf.

'As a matter of fact I've been in Central Asia,' I said. 'I crossed the Oxus.' What does it matter how I crossed it, I thought; after all, I had crossed it.

'Ah,' said Mr Alexis, working away. 'How very interesting. Coming from Kabul, I suppose, or going there? Where did you cross it?'

'I landed at Termez.' It was true in one sense. I had in fact flown over it, not the same thing at all, in the exploring sense, as crossing it by boat; but I was unnerved by this interrogation by a hairdresser and by the loss of my hair, which felt as if it were becoming non-existent under his ministrations; 'felt' because he would not let me look in the mirror while the work was in progress. In short, he had me on the run.

'At Termez. How wide would you say the river was *when you were there?*' The menace in his voice was unmistakeable.

'About half a mile.'

'You think as little as that? I suppose you crossed on a raft?'

This was the great divide. The moment when I began to lie.

'Yes, but of course there was quite a lot of other shipping about, paddle-steamers and so on.' I had seen the steamers from the plane.

I was telling *him*. What a fool I was making of myself. Of course, I should have known, with a name like that he was probably born in Russian Asia. Perhaps he had been one of our men in Termez. Somehow I had to put an end to this interrogation.

'Have *you* been in Termez?' I asked. It was about as profitable as addressing the Sphinx.

'Very dusty on the Russian bank. Didn't you find it so, sir? And Termez; really, just a single street.'

This was a chance to redeem myself. I took it.

'But of course on the Afghan side it's thick jungle in places,' I said. 'And they say' (who, I wondered, had said it, if anyone?) 'that there are tigers.'

He didn't seem to hear. He was doing something to the place where my double-crown had been when I had one.

'Used to be tigers,' he said at length. 'Certainly there *used* to be tigers.' And then, 'Of course you went to Merv?'

'No, but I flew over Samarkand.' 'Over Samarkand', it sounded feeble.

'Oh, you *flew?*'

'Yes, in a Russian plane, from Termez to Tashkent, then on to Moscow and from there to Vienna by way of Lvov and Budapest.' That will settle him, I thought.

'*Most* people fly from Kabul,' he said. 'I can't understand why *you* joined the plane at Termez.'

When he had finished I had the sort of hair-do that made me look as if I should have been driving a chariot. There was no need to comb it. What there was of it, until I washed it, just stayed put. It was hygienic and a bit cold at first. I liked it but I was not so sure that I was altogether happy at having the head of the Secret Service (Central Asia) as my barber, so after this I let Wanda cut my hair.

20

The Most Unforgettable Character
I Never Met
(1958)

On my return to London in the autumn of 1956 I had other problems besides thinning hair. I now had to write a book about our travels in Nuristan and keep my family alive while doing so. How I was to do this it was difficult to imagine, as the advance I had received from Secker had already been consumed. A possible source of income might have been the royalties from *The Last Grain Race*, for which I had received the lordly advance of £25 ($56), but as the book had been published only a couple of months previously I had experienced some difficulty in getting my hands on any of them, although I succeeded in doing so after Christmas. So while working on *A Short Walk in the Hindu Kush* I wrote articles for magazines as diverse as *Vogue* and *Everybody's* (now defunct), and articles and a number of short stories which Charles Wintour, editor of the *Evening Standard*, commissioned at £25 ($70) a time (I once wrote three in a week). Therefore it was fortunate that, at this critical moment in my life, the fortune that has so far never deserted me for long came to my aid when the chairman of Secker & Warburg, Fred Warburg, offered me a job promoting their products, in succession to Norman St John Stevas, who was leaving the firm for other fields (he is an MP and a former cabinet minister). I worked for Secker & Warburg until 1959 when, in the hope of earning more money, I left them to join the John Lewis Partnership, thus terminating what Fred was kind enough to describe as 'an hilarious association in publishing' on the flyleaf of my copy of his autobiography.

During these years, and also while I was working for the Partnership, I was able to supplement my earnings, which were pretty frugal, by writing long pieces for *Holiday*, an American travel magazine which had an enormous circulation. Although

Holiday treated its writers, many of whom were extremely distinguished (one of its most distinguished English contributors was V. S. Pritchett), rather as sultans treated their Circassian slaves, they could pay sums undreamed of in Britain to anyone with the strength and resilience to put up with their foibles.

One example of their rather unbending attitude towards their contributors, even in the face of adversity, will suffice. On one occasion, marooned in a remote hill village in Italy without a motor car and writing a very long article for them with an immutable deadline to meet, the typewriter key controlling the use of the capital letters broke. As there was no other typewriter in the village and no one capable of repairing one, I was forced to type a large part of the piece without employing any capitals at all. When the time came to dispatch the article to New York I wrote a letter explaining to the editor in charge the reason why I had been forced to act as I had done. By return I received from him an extended cable rebuking me for having sent him a manuscript in a form that I knew contravened their house rules. After this, wherever I went to work for *Holiday*, I always took two typewriters.

When the time came for *A Short Walk in the Hindu Kush* to be printed in the United States the publishers, Doubleday, began to suffer from the American equivalent of cold feet, in spite of its favourable reception in Britain, and they asked me to find some well-known person, preferably an authority on Central Asia, who would write an introduction explaining what the book was about and telling the Americans how good it was. According to them, it threatened to be unintelligible to anyone except a handful of reviewers and readers, not because of the lofty intellectual plane on which it was written but because it was 'too English', which made one wonder why they had bought it in the first place.

For a while I seriously considered the possibility of inventing a savant and writing the introduction myself, rather than approach some real authority who might not approve of what he might conceivably construe as a send-up of the kind of 'real' expedition which gets its supplies and funds by canvassing industrialists and such. The sort of man I had in mind would be a British Orientalist with a wartime record of service in some special, secret force; or else he would be older and his last expedition, a highly confidential mission, would have been undertaken at the behest of Lord Curzon while he was Viceroy of India. Neither of these men, for obvious reasons, would be a Fellow of the Royal Geographical Society, a member of the Royal Central Asian Society, or belong to any known club. Both would have refused knighthoods and any other sort of honour except those conferred by foreign universities which enabled either one of them to be referred to as 'Professor' if necessary; and both, for equally obvious reasons, would have always refused to allow their biographies to appear in *Who's Who*. Such a fabrication would not stand a hope in hell of succeeding in Britain but it might just have a chance of passing muster in the United States.

More sensibly, my thoughts turned to Evelyn Waugh, himself a very considerable and experienced traveller, with whom I had already been in correspondence about the book.

I had never met him. Almost everything I knew about him stemmed from reading his books. During the war we were often very close to one another: in the fog in Glasgow; in the same convoy in the Atlantic; in Cape Town where he did not find the girls as interesting as I had done; and in Egypt, though he had already left to return to Britain by the time I got there from India and where I frequently used to hear of something outrageous and funny that he had said. 'I do not propose to alter the habits of a lifetime to suit your temporary convenience,' was a much-quoted remark he was reputed to have made to a senior officer who had rebuked him for having gone on leave without permission. This, and the contents of half a dozen letters or postcards he wrote me from 1958 onwards, is the sum total of my acquaintanceship with Evelyn Waugh.

Our correspondence began when Secker & Warburg sent him an unbound proof copy of *A Short Walk*, in what must have seemed to them the faint hope of eliciting what is known in the trade as a 'quote' for use on the jacket.

Nothing, I thought, would be more likely to enrage him. But after a few days a meticulously written postcard arrived addressed to me, with a splendid quote and permission to cut or tinker with it if necessary. There was no need to do so, and anyway neither I nor the publishers would have dared. I wrote him a carefully worded letter of thanks in the legible calligraphic script I was then cultivating in order to set a good example to my children. Almost by return I received a reply.

The letter heading, Combe Florey House, his residence near Taunton, Somerset, was stamped in such grotesque black Gothic lettering that it looked like an invoice from a nineteenth-century undertaker's establishment. He was very kind, confessing that up to then he had confused me with P. H. Newby, 'whose works I have long relished' and who at that time was Head of the BBC Third Programme and the author of *Picnic at Sakhara*. Then, after going on about my book in a complimentary way for a bit and noting some slanginess and errors in syntax, he suddenly, with barely disguised malevolence, asked me how I knew that Wilfred Thesiger – the famous explorer whom I had met on his way into Nuristan while I was tottering, half-dead, out of it – was wearing, as I had written, 'an old tweed jacket of the sort worn by Eton boys' – implying that I could not have been privy to such information unless I had been there myself; and I could almost hear a triumphal snort of the kind that Charles Ryder's father utters in *Brideshead Revisited* when he suggests that his son, who has come down penniless from Oxford for the summer vacation, should attempt to negotiate a loan from the Jews in Jermyn Street to tide him over.

I had not been at Eton, but neither had Waugh. Why had he bothered to find out and how? I did not realize then, not yet having read *The Ordeal of Gilbert Pinfold*, that there was a distinct possibility that phantom voices were already accusing him of wishing that he had been at Eton, if not of actually pretending to have been at Eton himself.

I was anxious to propitiate him but reluctant to grovel. I really had seen Eton boys wearing such jackets on the playing-fields. Before replying, to make doubly sure that I had not been mistaken, I got in touch with Thesiger who was spending one of his rare periods in civilization, and asked him about it.

'Certainly,' Thesiger said. 'You're quite right. It's my old "change coat" from Billings & Edmonds.' Billings & Edmonds is still a well-known school outfitters in the West End.

Thinking what a gigantically impressive schoolboy Thesiger must have been at Eton if he was still wearing the same coat at the age of forty-five, I wrote to Waugh telling him, among other things, what Thesiger had said about his old 'change coat' and that I had seen change coats being worn on the playing-fields of Eton from the top of a bus while travelling from Slough to Windsor at the same time placing my dilemma about the introduction before him. It was not true – about the bus – but it was the simplest explanation I could think of and, now that Eton High Street has been closed to traffic, it is presumably an impossible feat. In the same letter and with the same kind of suicidal gesture to which Basil Seal, one of his own favourite creations, was particularly prone, I intimated that there was no such wine as the Clos de Bère of 1904, which, in *Brideshead Revisited*, Waugh allows Rex Mottram to give to Charles Ryder when they dine together at Paillard's in Paris. Didn't he mean Clos de Bèze?

I received the following postcard in reply:

Combe Florey House,
Nr. Taunton
1 August 1958

Thanks awfully for Grain Race [*The Last Grain Race*, of which I had sent him a copy] and interesting information about the other Newby. No book needs an introduction less than *Short Walk*, which is self-explanatory, I suppose the Americans want some certificate of bona fides. (When I thought you were the Egyptian don [P. H. Newby] I thought that perhaps the whole work was one of parody on a very high level.) I should get some explorer-mountaineer to introduce you. I cut no ice in America nowadays. Bère is a misprint that follows that book into every edition. But if all explorers fail, I *would* write an introduction.

E.W.

Waugh then wrote the following postcard to A. D. Peters, his literary agent:

Combe Florey House
6 August 1958

Could you kindly make a discreet enquiry for me? A Mr
Eric Newby has written an excellent book about a trip to
Afghanistan called A Short Walk in the Hindu Kush, to be
published by Secker & Warburg. He tells me that
Doubleday in USA will only publish it if it has an
introduction by someone else. If this is true, I am willing to
do it, but if he is pulling my leg and merely wanting an
extra puff for his book, I won't. Terms don't matter. I
should be doing it as a kindness.

E.

A week or so passed and then on 16 August a postcard arrived:

15 August 1958

I have written a short preface to A Short Walk and sent it
off. I don't know if it will be what the Americans want. I
have lost touch with them for the last ten years. Anyway, I
was glad to try.

E.W.

The preface was a splendid piece, spirited but also rather sad,
written, as Waugh said, by a man 'whose travelling days are
done'. It was plain that before writing it he had also taken the
trouble to read *The Last Grain Race*. How was I to repay this act of
kindness? With money? But how much money did you offer
someone of the fame and stature of Evelyn Waugh or, worse still,
to his agent? It seemed an insoluble problem. Then I thought of
the Clos de Bère/Bèze. Why not send him half a dozen bottles,
two or three magnums if such existed, of this splendid burgundy
– not of course the 1904 which, even if it was still drinkable in
1958, would be beyond my means – with my thanks and a note
to the effect that the label, which my wine merchants would
either reprint or alter to read Clos de Bère, was a case of nature
imitating art?

The wine merchants were not at all enthusiastic about my

171

plan. In fact they were depressingly sycophantic in their attitude to Mr Waugh, even to Mr Waugh *in absentia*. They did not have the pleasure of serving him, they said, implying that they would, however, welcome the opportunity of doing so; and they would certainly not lend their names to the sort of deception I was proposing. In the end I had to content myself with sending him three magnums of Clos de Bèze, correctly labelled. I forget the year. At the time it did not occur to me to alter the spelling on the labels myself or to commission an art student to produce some bogus ones, perhaps because it meant that I would have to pack the bottles up and send them off to Waugh myself.

> Combe Florey House,
> Combe Florey,
> Taunton
> 31 August 1938 [sic]

My dear Newby,

I don't often get presents nowadays. I have seldom ever had one as splendid as the three magnums of burgundy which arrived yesterday. *It is not a wine I have ever tasted* [my italics] but I know enough to realize that there is an enormous pleasure ahead of me – three enormous pleasures.

I have reverently laid the bottles in the cellar and am resolved to leave them there until my 58th, 59th and 60th birthdays, by which time I am confident that you will be solidly established as a writer and justly admired.

Thank you also for the charming verses which I never saw before. I hope that if by any lucky chance you should be in Somerset in late October 1961, –2 or –3 you will drink the wine with me.

> Yours sincerely,
> Evelyn Waugh

I replied that I would be happy to do so, but I never pursued the matter and had no intention of going, even if I was invited, and I am sure if he had known this he would have applauded my decision. It would have spoiled our relationship if we had met. Nevertheless, in 1959, when my publishers asked him for his views on *Something Wholesale*, the book about my life in the rag trade, he sent them another enthusiastic quote to use on the jacket, unasked. I hope the wine was good if and when he eventually opened it.

Altogether I saw him four times. Once was in March 1957 at the Law Courts, which curiosity impelled me to visit during his fantastic lawsuit with the *Daily Express* and one of their columnists, Nancy Spain. This lawsuit was brought about by Miss Spain's ill-advised attempt, together with Lord Noel-Buxton, to enter Waugh's house uninvited, which led to her writing things about him which were libellous. I am not sure on the day I was there whether or not he was wearing a false moustache. I am slightly more sure that he was using an ear-trumpet. Spain herself was wearing one of the nearly full-length mink coats that the Beaverbrook Press used to keep in mothballs for their female staff to wear when appearing on such occasions, but over trousers, which was eccentric then. He was awarded £2000 ($5580) damages plus costs, and the *Daily Express* settled out of court for a further £3000 ($8370) when Spain continued her hostile comments in another context, which also extended to Graham Greene.

On another occasion I saw him on the steps of White's Club, at the top of St James's Street, trying to hail a taxi. Apart from the fact that by then he was much older and wearing a fearsomely checked suit, he was exactly as he appears in a caricature by Osbert Lancaster, done in 1947, which he himself bought in order, so he said, to frighten his children.

The third time he nearly bowled me over turning into King Street from St James's Street, probably en route for Christie's or the London Library. That time I think perhaps he recognized me, or possibly he thought I was something from his Pinfoldian past. He not only looked at me intently but turned round after he had passed for a further look, just as I did, so that we were like the two Englishmen in Kinglake's *Eothen* who passed one another in the Sinai Desert in the 1840s. The only difference was that they were mounted on camels, and they did eventually retrace their steps to speak to one another. We, on the other hand, continued on our different courses.

The last time was one afternoon when I was having tea *en*

famille at the Ritz. Seated in a corner under a niche containing a piece of statuary, he was dressed in one of his indestructible, immensely checked suits and eating buttered toast (or it might have been muffins), it was difficult to see through the murk in that vast room on an autumn afternoon. He looked exactly like a drawing by Phiz in *Mr Sponge's Sporting Tour*. 'Go on,' said my children, when I told them who it was, 'speak to him.'

I would have liked to but I was afraid that when I introduced myself he might say, 'And what does that mean to me?', as he could well have made one of his characters say, such as Ivor Claire, the idle, elegant Captain of the Blues (The Royal Horse Guards) – in *Officers and Gentlemen*, the second novel in Waugh's wartime trilogy – who deserts his men in Crete and is later saved from disgrace by the Establishment and shipped off to the East as a kinsman of the Viceroy, where he wins the DSO in Burma. Claire's particular butt in the commando of which he was a member was Trimmer, a rather caddish, one-time barber who, through no fault of his own, gains an unmerited decoration, and becomes an undeserving hero. Like Trimmer, I had been an Englishman in a Highland regiment; like Trimmer, I had gained an unmerited decoration. But, unlike Trimmer, I did not have the constitution to accept such a rebuff if it was offered, although I admired Waugh's writing and I was extremely grateful to him for what he had done for me. So I did not speak to him.

21

A Visitor from Lhasa
(1958)

The only incident worth recording here, hilarious as much but by no means all the time I spent with Secker & Warburg undoubtedly was, is one during which I had my first and probably last encounter with a Tibetan Lama. This took place in Southern Ireland in the course of the unravelling, or rather an attempt by numbers of people more or less competent to do so, to unravel what came to be known as 'The Great Lama Mystery'.

In 1955 Fred Warburg was instrumental in acquiring from an agent the autobiography of T. Lobsang Rampa, a Tibetan Lama, entitled *The Third Eye*. On publication it became a tremendous and instant success, receiving enthusiastic notices from the most eminent reviewers, although some of the more erudite were frankly sceptical; and indeed – whether it was true or not – it was a most vividly written and fascinating work. It was subsequently published in numbers of other countries and its sales were, and continued to be after the events which I now propose to describe, enormous. The real mystery was the Lama himself, who had adopted the name of Dr Kuan-suo. Although photographs of him existed and were in fact used for publicity purposes, few people apart from publishers and literary agents claimed to have met him, although he was reputed to be resident in Britain. Then, at the moment when the Lama's book was at the height of its success at the end of January 1958, the *Daily Mail* fired off what was to prove a bombshell. It published an article which said that Lobsang Rampa was not a Lama. And not only was he not a Lama but he was, in fact, Cyril Henry Hoskins, son of a plumber from Plympton in Devonshire, married to a lady who had been a state registered nurse in Richmond. If what the *Mail* said was true, the future of *The Third Eye*, and the Lama with it, looked bleak.

On the Sunday following the day on which the affairs of the Lama had reached this stage of dénouement, I was at home in Wimbledon trying to get on with a book (which in fact meant that I was just sitting, feeling terribly cold and wondering if we could all emigrate), when the telephone rang. It was Fred, asking or rather ordering me to go to Ireland at once. Apparently the Lama had fled the country and had taken refuge in a house near Howth, a seaside resort on a peninsula on the north side of Dublin Bay. There he had been discovered by the press of the world, who had surrounded the building, and he was now barricaded inside it. My job, Fred said, was to gain admittance to the building and ask the Lama for an explanation. He was unable to go himself as he had promised to take part in a broadcast from Birmingham the following day in which he was to cross swords with the Secretary of the Public Morality Council, or some such body.

Secretly I was delighted to be entrusted with this mission, which was just the thing to break the tedium and, after collecting my passport, which proved to be unnecessary, and putting on a sheepskin coat and a pair of sheepskin boots – it seemed likely to be pretty cold taking part in a vigil on a headland near Howth – I went to Fred's flat in St John's Wood where he gave me some money and a letter of introduction to the Lama, the gist of which was that I was acting as an emissary on behalf of the publishers and that he must make some statement about himself and his book.

By this time the *Daily Express* had been on the line to Warburg. They were furious at having been scooped by the *Daily Mail* and were anxious, if it was at all possible, to prove that the Lama was a Lama and not the son of a plumber in Plympton. They wanted one of their reporters to accompany me to Ireland for this purpose and Warburg thought this would be no bad thing. At such a moment it must have seemed to him that any allies were better than none.

At the *Express* office I was given a large sum in £5 notes for my expenses, which I was never asked to account for and which was a great help with the next term's school fees. By the time we had flown to Dublin and had reached Howth it was much too late to call on a Lama or anyone else, and I spent much of the rest of the night listening to the reporter telling me about life on the *Daily Express* whom, he said, were generous employers, especially in times of adversity; a sentiment with which, with a great load of £5 notes in my pocket, I thoroughly agreed.

We both woke the following morning with terrible headaches and quite early set off for the Lama's house, which was situated on the edge of a cliff with a stupendous view over the sea. It was a medium-sized villa surrounded by a wall and also by representatives of the press of various national and other newspapers.

I had never seen the gentlemen of the press in action before – and frustrated gentlemen at that – and now doing so I found it difficult to believe my eyes. Some of them had constructed primitive periscopes, using bamboo poles with mirrors attached to them, with which they were trying to look in through the windows of the upper floor, presumably trying to discover what a Lama looked like in his bath. Others were apparently happily engaged in going through the dustbins to see if they could find any interesting material, and the approaches to the house were littered with paper and other rubbish which they had rejected. When I arrived, this body of persons rather surprisingly fell back to allow me to get as far as the front door.

After repeated bangings on it Madam Kuan – it sounded like Kwamph – the wife of the Lama, came to the door but refused to open it, and we had a difficult conversation through the letter-box, the gist of which was that I should come back later.

When I came back later I was allowed in to the hall of the house, which was furnished with a wooden Buddha and a brass tray made in Birmingham on a rickety black ebony stand. It was a depressing room, rather like the set for an oriental interior in a play to be performed in a village hall. Madam K. herself was plumpish, middle-aged with greying hair done up in a bun, vaguely foreign-looking. With her was a fresh-faced, very English-looking girl who told me that she had left her husband (who was a member of Lloyds), and her three children in order to live as a disciple in the Lama's house. She was 'upset' at being, like the Lama, subjected to a lot of undesirable publicity.

The Lama, Madam K. said, was dying. He had suffered a coronary thrombosis and it was unlikely that he would live for more than a couple of days. Nevertheless, he felt that he should make a statement, and would I come back the following morning.

I passed another dissipated night, this time with a whole band of journalists, and the following morning went back to the house, to which no one had up to now succeeded in gaining admission except myself.

Once again I found myself in the hall, beyond which I was not allowed to penetrate, and in it, without seeing the Lama, I listened to two tape recordings which he and his wife had made, apparently for my and Warburg's benefit. The Lama's statement, which was remarkable for its fluency, described how Cyril Henry Hoskins had become a Lama. At the age of about thirty-four Hoskins received advanced warning, presumably by some process analogous to mental telepathy, that his corporeal integuments were going to be taken over by a Tibetan Lama and that the actual take-over would occur on 13 June 1949. He then went on to give an account of the events leading up to the moment of actual incorporation. Some time during the latter part of the war, it must have been in 1944, the Lama, whose name was Lobsang Rampa, left Lhasa and travelled to Chunking where he worked in a displaced persons camp as a doctor (he had a medical degree). While he was there he performed a number of major operations, some of them on American and English women, who were among the inmates. Eventually he became a prisoner of the Japanese and was in Hiroshima when the bomb exploded. In the resulting chaos he stole a fishing-boat and managed to reach Vladivostock by way of Korea. He then crossed Asiatic Russia by the Trans-Siberian Railway, some of the time hanging to the tie-rods beneath the flat-cars and supplementing a starvation diet by eating the grease out of the axle-boxes. Eventually he arrived in Moscow; there he was arrested and taken to the Russian-Polish frontier where he was deposited on the Polish side of it. From Poland he travelled to Germany where he took a job as a worker in what I think he said was the Daimler-Benz factory. From Germany he made his way to England and to Woking for his tryst with Hoskins who, at that time, was living in a house opposite the Cottage Hospital.

Hoskins was keen on photography and on that particular day
– 13 June 1949 – he was up a tree trying to photograph an owl.
While endeavouring to obtain a close-up the branch on which
he was extended broke, and he fell. The statement gave a vivid
account of this fall. How green the lawn looked and how many
ladybirds and other insects he could see walking up and down
the blades of grass as the earth zoomed up to meet him and he hit
the ground with a terrible thud.

The next thing he knew was that he was outside his body,
looking down at his own supine form to which, advancing across
the grass, but with its feet just clear of it, was a Lama dressed in
saffron robes. When the Lama was quite close to Hoskins he said,
'I want your body.'

'Why?' Hoskins asked.

'Because I haven't got a passport and if I don't have it I shall
be deported,' the Lama said.

At this moment, the one which presumably Hoskins had been
awaiting, he made an act of acceptance and the Lama passed into
the body of Cyril Henry Hoskins and possessed it. And from this
time on Hoskins was the Lama and Hoskins was no more.

I took the tapes but I told Madam K. that I was not going to
leave Howth until I had seen the Lama. And on the afternoon of
that day I was admitted to the room in which the Lama was
supposedly dying. It, too, was sparsely furnished: a bed in which
the Lama was tucked up with only his head visible, a couple of
chairs and a popular encyclopaedia in one corner. The view was
splendid over the water to Ireland's Eye, an island just off the
coast. The wind was easterly and very cold and the sea was
smooth. The Lama's face made a powerful impression on me.
He had a long nose, a beard which made it difficult to see his
mouth and teeth, and rather red lips. His head was bald on top
and he had a high, domed forehead in which there was a slight
dent which could have been the result of the 'third eye'
operation he described in his book, or could equally well have
had some more mundane cause. But his eyes were his most
impressive feature. They were very luminous and powerful.
Altogether the whole feeling the Lama gave me was one of great
power and although it may sound imaginative, I could not help
thinking of Rasputin and how difficult it had been for his

assassins actually to kill him. What I was looking at was a man who looked as if he, too, would resist death strongly. He certainly didn't look as if he was on the point of it, although he didn't look particularly well, and next to him on a side table there was a hypodermic syringe with one or two spare needles on a tray.

The Lama's breath was coming in a halting, gasping fashion.

'I've been taken for a ride,' was the first thing he said to me, enunciating with difficulty. 'Now go back and tell Fred that the next ride I'll take will be in a box with glass sides.'

He had a very strange voice. If the accent was Devonian it had other, more urban overtones. It certainly wasn't oriental in any way, neither were the expressions which came from his lips those of an oriental person.

'I never took any money from anybody,' was the next thing he said. (It had been suggested that he had taken money more or less under false pretences for horoscopes, advice, things like that; but one of the strange things about the whole affair was that, although these suggestions were given wide publicity, no one ever came forward, as they might have been expected to do, to say that he or she had sent a three-and-sixpenny postal order with a stamped, addressed envelope to the Lama and got nothing in return.)

Then he said, 'The book is a true book,' and continued to do so during the entire time I spent with him. And when I asked him to autograph a copy of it for me he wrote 'This is a true book' in it and signed it 'T. Lobsang Rampa'. And with this I had to be content. He implied that he was dying. His wife said that he was dying, he obviously wanted me to think that he was dying and it was impossible to harry a man in that condition. So, having expressed the hope that he would soon recover, I left.

The Lama's recovery was far more rapid than I would ever have dreamed possible. Madam K., convinced by the reporter from the *Daily Express*, who by now had a staff photographer from the Manchester office with him, that his paper was intent on displaying the Lama in more favourable light than he had been by the *Daily Mail*, had admitted them to the house, and as I left they were preparing to enter what they, too, had been led to believe was the death chamber.

What then happened was this. As soon as they went in, the Lama allowed Madam K. to prop him up with some cushions and he then had a number of photographs taken of himself with a cat perched on one shoulder. What the journalist wrote about this encounter I cannot now remember. What I can remember is being back in the Shelbourne Hotel in Dublin and the photographer, a cheerful, extrovert sort of fellow who would have been equally at home at a post-mortem or a Royal garden party, going off somewhere to get his pictures developed. The next thing I remember he was on the telephone to us in the hotel where we were wondering what to do next, with the words: 'You know those pictures I took of the Lama? Well, I've printed them oop and do you know what? There's a ruddy great 'alo round 'is 'ead!' And indeed subsequent examination did demonstrate that there was a certain phosphorescence around it.

For months and years afterwards letters continued to arrive at the London offices of Secker & Warburg, either for onward transmission to the Lama or concerning him. With the exception of those who had actually been in Tibet or Central Asia, the majority of writers were not interested in whether Hoskins was a Lama or the Lama was Hoskins. They wanted to believe in this strange traveller who had come to the West from what had been for most of them an unknown world. Some of these letters were expressed on a high level of fantasy. One of them, sent by a M. Morkeh who lived on the Côte d'Ivoire in French West Africa and addressed to Fred Warburg, apparently under the impression that Fred and the Lama were one and the same, read as follows:

Dear Sir,

Many thanks to God the Almighty that He has given you to me. I know your place is the light, and my sufferings in seeking for the spiritual light will soon be ended. Now, I have spent a lot of money in so many societies to open my Third Eye of the Soul Eye, but I am still in darkness. For many people are using false teachings for the purpose of their simple living. Therefore, I wish to put an end to every society, and have great correspondences with you till the end.' ['Oh, no, he's not,' Fred said, firmly, when I read this to him.] 'I believe that by walking with you for some time, my Soul's Eye will be opened. Immediately you receive this my humble letter, try to send me your list or a catalogue, if you have some books to borrow, state to me. Try to send me one strong prayer book of help. Awaiting you for helpful, favourable and immediate reply by the return airmail.

As with other such letters, if any such can be imagined, there was very little that could be done as neither Fred, nor apparently the Lama's agent, had any idea of where, by this time, the Lama had gone.

22

MG Buyer
(1960–3)

I joined the John Lewis Partnership at the invitation of the Chairman in the autumn of 1959. The Partnership was, and still is, a highly successful and efficient business organization, the members of which, known as Partners, own, among other ventures, a chain of department stores. Originally, when I first talked with him, his idea was that I might be employed in some part of the organization in which my literary capabilities could be used, perhaps in a section dealing with public relations or internal propaganda, which played an important part in making the Partners credible to themselves and in generally reinforcing the Partnership's mystique; but he later came to the conclusion that initially I ought to be employed in the commercial side.

That autumn I became a shop assistant in the China and Glass Department at Peter Jones in Sloane Square, the most sophisticated of the Partnership's stores, just in time to take part in the Christmas rush. Apart from working in Piece Goods, or Carpets, which also required a considerable degree of physical fitness, China and Glass was one of the tougher assignments, from the point of view of an assistant, that one could have at Peter Jones. What were on show in the department – which was bang opposite the main entrance on the ground floor – in the way of tea sets and dinner services were mostly prototypes, the majority of the stock being kept in the basement. This meant that every time anyone made a sale of a tea set, for instance, it was necessary to go below and assemble one in the stockroom from its constituent parts. By the time the store shut in the evening, particularly during the weeks before Christmas when it kept open late, after innumerable descents and ascents to and from the stockroom, one was on one's knees.

On my first day in China and Glass I made two spectacular

sales which seemed to augur well for my future success in this field. The first, ten minutes after the store had opened on Monday morning (by which time I had not even mastered the technique of making out a bill), was to a beautiful Canadian widow, dressed in funeral black which, if anything, added to her charms. It was for an enormous dinner service hand-painted with reproductions of Thornton's illustrations for his *Temple of Flora*, which cost what was at that time a small fortune, and for which she actually wanted to pay in advance.

This dinner service, the produce of an illustrious firm in the Potteries, was only made to special order, and in those happy, far-off days in which demand outstripped supply, took anything up to a year to complete. In this particular instance it took so long that, by the time the dinner service had been despatched to Peter Jones from the factory, the beautiful widow had herself died and it had to be delivered to her executors.

My second sale was of an extremely expensive, and I would have thought highly undesirable, piece of statuary executed in Lalique glass. I made it to a lady from St John's Wood with rather alarmingly tinted hair, which made her look as if she had a tortoise perched on her head. These two sales, which had an electrifying effect on the department's weekly sales statistics in what were the pre-Christmas doldrums, were regarded with some awe by everyone concerned, not least myself. Unfortunately they proved to be nothing more than beginner's luck, and for the rest of my time in the department I hardly ever sold anything that retailed for more than £20 ($56).

From China and Glass I was sent to the Do-It-Yourself Department, at that time a comparative innovation, in Jones Brothers, a cheerful, less classy Partnership store in the Holloway Road, where I sold tintacks, nails, hinges, glue and so on. From it I returned to Peter Jones, where I became an assistant in Women's Coats – I was debarred from working in the Dress Department as I would not have been allowed in the fitting rooms. I found that I had a considerable aptitude for this work, which was not really surprising since I had spent so much of my life selling coats to store buyers. Some of the coats I sold were extremely expensive models from French and Italian couture houses that had been bought for copying purposes; they were then put on sale and knocked down to the customers for what, although they did not appreciate the fact, was a fraction of what they originally cost. There were regulations in the Partnership regarding the marking-down of what was known as age-rated stock; these regulations were so strictly enforced that when the Partnership first began to offer antique silver to its customers at what was already a highly competitive price and it failed to sell within the time limit allowed, it too became age-rated and was marked down, at which moment it was snapped up by dealers in the King's Road for profitable re-sale.

From the Coat Department I went to the Intelligence Department, housed in what was then the new John Lewis store in Oxford Street, where I was put to work on Fashion. One of the main tasks of the Intelligence Department, which was widely feared by the Partnership's central buyers (as I was to fear it when I, too, became a buyer), was to evaluate the stocks held in its own stores and in those of its competitors. It also kept a continuous check on the prices its competitors charged, in support of the Partnership's rather boastful-sounding claim, which was perfectly genuine although few of its customers believed it, that it was 'Never Knowingly Undersold'. In fact any Partner who successfully identified an undersale on a competitor's premises was rewarded, and some of the younger and fitter assistants in Oxford Street at that time used to supplement their wages by doing this during their lunch-hour.

As a member of Intelligence I spent hours going through the rails of our own stores, which were strung out between Southampton and Liverpool, and those of rival stores and fashion shops. While doing so I would record the entire order of battle of some particular department such as, for example, wedding dresses, noting what materials they were made of, how many were in stock of each, and in what sizes, and whenever possible I would identify the manufacturer. You could often do this by simply asking an assistant.

While I was working for Intelligence I was offered what everyone said was one of the best buyerships in the Partnership, that of women's stockings. I was aghast when I discovered that in order to learn enough about stockings to be able to fill this post I would have to spend some months behind the counter at Peter Jones and at one of the provincial stores, selling stockings to the customers. It was not so much the idea of buying stockings that I minded, although without legs in them it was difficult to imagine anything more boring than these deflated by-products of lumps of coal. It was simply that I had never seen a man selling stockings in a department store and I did not fancy being a pioneer in this field.

I subsequently received a letter from the Director of Trading (Department Stores), the gist of which was that in the course of our reviewing together at some length the advantages and disadvantages of my becoming the Central Buyer of the Stocking Department, I had made it clear that while perfectly willing to do whatever the Partnership asked I had at present a distaste for Stockings as a commodity. He, on the other hand thought the suggested appointment was the right thing for the Partnership and myself and that none of the management thought it likely that I would be required to hold this particular Buyership for a long term of years. All buying was interesting, and over much of the buying field the mechanics of the operation were as interesting as the merchandise purchased, and if I really looked this problem in the face, I would accept the Buyership . . .

I did what he suggested – looked women's stockings in the face – and accepted the Buyership. I was constrained to do so by a remark that the Director of Trading let fall in the course of our conversation, in reply to my question as to what would happen if I did not accept. 'Shall we say,' he said, 'that your future use to the Partnership might be somewhat limited.'

On the following Monday morning, having had a manicure in order to avoid ruining the merchandise, I reported for duty in the Stocking Department at Peter Jones. There, confronted by my first customer, I thought I was going to faint. No words would come out in reply to her request for a pair of sixty gauge, fifteen denier, seamless supersheers in misty beige, I think it was. I slunk away and telephoned the Director of Personnel and told him that I could not be the Stocking Buyer. If necessary I would resign from the Partnership. He thought I was queer, as everyone else did, but he told me to go back to the Intelligence

Department and continue my work there, instead, while my future, if any, with the Partnership was considered.

Back with Intelligence, which was entirely staffed by women Partners at that time, I was given a gigantic task, one which unless they were transvestites they could scarcely carry out themselves. I was ordered to evaluate the stocks of men's suits, odd jackets, blazers, trousers, and raincoats in six of the Partnership's stores and in those of their rivals, testing them for value, sizing (I was a stock size 4 long) and fit. This meant, in addidion to trying on raincoats, jackets and blazers in our own and rival establishments, that I was undressing more or less completely anything up to forty or fifty times a day. Before setting out I was given money in case some super-salesman got me into a position when I had to buy one of his suits, but I never did. By the time this lengthy investigation drew to a close I had lost most of the fly-buttons on two of my own suits.

The last city in which I carried out my investigations was Southampton and, while returning to London on an evening train, I was approached by a small man who introduced himself to me as the Assistant General Inspector of the John Lewis Partnership, an appellation which made him sound like something out of Kafka rather than someone who had worked for a German stores group until Hitler appeared on the scene.

'I know what you are doing,' this individual said, the only one in the Partnership I had so far idendified as having his shirts made by Turnbull & Asser and getting his hats from Lock. 'And now I want to ask you what, in your opinion, is the best medium-weight woollen suit retailing for less than fifteen pounds [$42] that you have discovered in the course of your evaluations?'

'I think,' I said, since my time in the Intelligence Department had made me almost as didactic as he was, 'that the best suit of this kind is marketed by C & A, in their store in Oxford Street. It is made in Holland in two shades, blue mixture and grey mixture. It is made in all normal sizes in short, regular, long, portly and regular portly. Even the foreparts of the legs of the trousers are lined, which is very rare in ready-to-wear clothing in this country, although it is commonplace in Italy. This suit is the best within the limits you set. I have no doubt of this. I can tell you more about it if you would like me to.'

'Good,' he said. 'I, too, am of the same opinion.'

Although I suspected that what he said was bullshit, I could not help but admire his nerve. Perhaps as a result of this confrontation and the huge report I wrote which embalmed the results of my findings on men's clothes, or perhaps as a result of various other confrontations, I was offered the Central Buyership of MG (otherwise Model Gowns), which in the Partnership at that time meant all dresses retailing at ten guineas and upwards, for twelve of their stores.

The position of MG Buyer was, according to the testimony of one or two of the previous incumbents to whom I had been able to speak who no longer worked for the Partnership, about as safe as working in a factory producing nitro-glycerine. However, before taking it up I worked for six months as assistant to Lionel Wharrad, the Managing Director of Peter Jones, a marvellously dynamic Welshman who had risen from the ranks and was adored by his staff. He was often at loggerheads with the Central Administration of the Partnership. My job was to provide him with facts and statistics that he could use to prove to them that so far as fashion was concerned at Peter Jones, Central Buying was less efficient than buying carried out by buyers who bought for Peter Jones alone, buyers who were, in fact, also department managers. This period, helping him to do battle with the Central Administration, was the only one I really enjoyed in the years I worked for the Partnership.

As the MG Buyer I bought day dresses, dresses with jackets, long and short evening dresses, what used to be called 'cocktail dresses', wedding dresses – the buying of which for hours on end filled me with the same sort of horror of spectral whiteness to which Melville devoted some space in *Moby Dick* – and bridesmaids' dresses, the most unbuyable and unwearable of all dresses.

I never realized how popular I could be until I became Central Buyer of dresses retailing at ten guineas (about $28) and upwards, and sat with my new-found friends, the manufacturers, in their showrooms in the wholesale jungles east and west of Great Portland Street, or in the equally ghastly regions of what had once been Mayfair, watching the model girls come prancing on like racehorses that had been given a shot of steroid in garments that Louie, or Harry or Charlie or Sidney told me were going to be 'very big'.

Talking to them I used to feel as if I had taken one of those courses by post – 'You, too, can be the life and soul of the party' – and that I was witty, gay, even astute (well, perhaps not astute) and even, on occasions, funny. 'Now this is going to be very big, Mr Newby, very very big' – this as some dreary old number zoomed past. 'Come back, Patsy, and let Mr Newby have another look at it . . . Very funny, Mr Newby. He's got a wonderful sense of humour, Mr Newby has, hasn't he, Joyce? Can I get you some sandwiches, Mr Newby? Some smoked salmon sandwiches for Mr Newby! And what about a noggin, Mr Newby? A glass of champagne for Mr Newby! I expect you'd like to get the order down now, Mr Newby, so I'll leave you with Joyce.'

Relatively little of a Central Buyer's time was taken up with actual buying. Much more of it was spent deciding how much of the total season's budget ought to be committed initially; how much of this to devote to each category – evening dresses, wedding dresses and so on – and how many of each size to buy.

Much of my time was spent travelling between Oxford Street and Brixton, Southampton, Sheffield, Liverpool, Sloane Square, Nottingham, Holloway, Reading, Hampstead, Cambridge, Southsea and Windsor, marking down slow-moving stock and discussing one's mistakes and successes, if any, with the department managers, those sterling women who, having been in wholesale myself, I felt I had known for most of my life. By the time this enormous round was completed, it was time to start all over again. I felt like one of the painters of the Forth Bridge.

Living in this way it was only too easy to forget to look and see what women were actually wearing in the field, as it were, and on one occasion my attempt to find out involved me in what promised to become a *cause célèbre*. This was when, with the knowledge of the Director of Buying, Fashions, my immediate boss, I paid a visit to Epsom and Ascot in order to find out what women were wearing at the races, the visit to Ascot taking place while on my way back to London from a visit to our branch at

Reading. Having done this I found that the Director of Buying, Fashions, was reluctant to countersign what were my comparatively modest expenses, which included the cost of entry to the paddock at both courses.

20 June 1961

Memorandum. Director of Buying, Fashions, from Central Buyer, MG Department.

Request that the Director of Buying, Fashions find out from whatever authority is appropriate whether or not expenses incurred in obtaining entrance to those parts of Ascot and Epsom Racecourse frequented by women who might be expected to be wearing outer garments costing more than ten guineas (retail) are chargeable to expenses when such money is spent by a member of the Partnership *wholly* with the intention of furthering his knowledge of such merchandise with a view to improving his buying of it in the future.

In reply the Director of Buying, Fashions, received a memorandum from the Director of Trading (Department Stores), headed Mr Newby's Expenses at Ascot and Epsom, his note to you of 20.6.61, to the effect that to the best of his knowledge no Partner had ever asked to be reimbursed for expenses incurred in going to Ascot and Epsom and into the enclosures, and he was doubtful whether such expenditure would prove to be allowable for tax in the Partnership's accounts. However, on the assumption that I had incurred this expense in the genuine expectation that the Partnership would be ready to reimburse me, he would be prepared to agree to that being done, but only on the understanding that it was not to be regarded as a precedent, and that no similar claim for reimbursement was to be made in the future unless prior agreement to the expenditure has been obtained.

26 June 1961

Memorandum. Director of Buying, Fashions, from Central Buyer, MG Department.

. . . in spite of the ruling of the Director of Trading (Department Stores) and in order that there should be no misunderstanding about the motive of the enquiry I have decided not to charge the expenses for the visits which I made to Ascot and Epsom and I would be glad if you will inform him that I am not doing so.

22 August 1961

Memorandum. Director of Trading (Department Stores) from Central Buyer, MG Department.

No. 217   Your Memorandum 2586

1. While understanding that it is necessary to record all the salient points in an interview such as the one that took place yesterday I do feel surprised that you should have recorded your remarks to me about Ascot in paragraph 7 of the above memorandum, particularly as this memorandum will take its place in my permanent dossier and might be seriously misinterpreted by someone who, unlike yourself, is not in full possession of the facts . . . [A reading of this paragraph might give the impression that I had been detected on a clandestine visit to a racecourse in the pursuit of pleasure in the Partnership's time, which was not the case.]

On 24 August 1961 the Cavendish Registrar received a memorandum from the Director of Trading (Department Stores) with reference to memorandum no. 217 written by me to him. The fact was that the visit to Ascot referred to had been made with the full knowledge of my Director of Buying, and that he thought the best and fairest way of meeting this point would be for his memorandum and mine to be filed with his earlier memorandum 2586.

This was not the end of the matter. It continued to rumble on and be referred to in other memoranda as if it had never come up for consideration before, and I realized that although I had won a battle I was in process of losing the war.

It was not only in the realms of the higher bureaucracy that I was running into danger. One of the advantages or disadvantages – according to what sort of temperament one possessed – of working for such a paternalistic, some might say maternalistic, organization as the Partnership was that its members were not only encouraged to engage in one or other of the extramural activities which it sponsored but, when the chips were down, were expected to do so. Thus, with the Partnership looking on benevolently from some ivory tower, if you were not careful you could easily find yourself playing football, learning to dance the carioca, sailing round a lighthouse and back, not with your 'partner in life's race' but with another sort of partner who, if you went on doing it, might quite easily usurp the position held by the real one. Even worse, you could find yourself engaged with them in developing photographs, playing the lead with them in *Blithe Spirit*, spending weekends with them in Partnership houses, even skiing with them in the Alps, although at that time invitations to the Alps were issued by the Chairman only and were normally only extended to those who showed promise of becoming exceptionally Partnership-minded. Apart from the skiing party, these pleasures were open to all Partners whatever their rank, and in fact if one indulged in even a small part of the activities available to Partners during one's time off, there was scarcely any need to go home at all.

But in accepting the position of MG Buyer I soon discovered that I had become a member of a hierarchy within the Partnership which comprised the upper crust, and one which was able to indulge itself with a minimum of expenditure in pastimes that I had previously thought of as only being available either to the very rich or else to landed members of the upper class. As a member of this hierarchy I could, if I wished to do so, spend my weekends in a splendid country house in Hampshire, one that was equipped with good cellars and with excellent pheasant shooting and some of the finest fishing in the world immediately outside its door. These were not the only pleasures available. In summer I could, and did until I got bored with it, go yacht racing in the Channel with other members of the upper hierarchy – none of the lower ones was ever present when I was a member of the crew. I later learned that because of my previous experience as a sailor it was particularly hoped that I would interest myself in the Yacht Club, the commodore of which was a member of the Royal Yacht Squadron, and when I failed to do so I knew (because someone in a position to know told me) that a black mark had been recorded against my name.

It was unfortunate that I was not interested in doing any of the things that the Partnership had dreamed up on my behalf and that of my fellow Partners, at least within the Partnership. My ideas of pleasure did not include drinking even the best of wines under the beady eyes of some director of something or other and his wife, either one of whom might make a mental note if one helped oneself too freely and might communicate this information to some higher authority, who might very well note it down and have it filed away in one's personal dossier. Nor did they include shooting with him or fishing with him or sailing with him. The last thing I wanted to see after the day's work was done was a Partner in any shape or form, and the truth was that I was not cut out to be one.

In order to gather inspiration at more rarefied levels than the rather mundane ones at which the nature of their tasks required them to operate, and also as a change from the interminable round of branch visiting, most of the fashion buyers used to go twice a year to France and sometimes to Italy in order to attend the showings of the couture and boutique collections there.

Among the stranger forms of prohibition which operated at that time within the Partnership was one which forbade buyers on Partnership business from travelling first-class by air, although it was perfectly permissible when going abroad to travel first-class by train. (Whether this prohibition extended to all members of the upper hierarchy, I do not know, nor do I care.)

As a result, the majority of fashion buyers (what the buyer of Oriental Rugs did, who might well have had to visit China, is not clear), thoroughly fed up with endlessly travelling round England in second-class railway carriages while visiting the branches, elected to travel first-class by train when they went abroad, justifying the extra cost of doing so with the words that all Partners had used since time immemorial when justifying some piece of conduct that was just teetering on the edge of being unacceptable: 'The Partnership would wish it.'

And it really was expensive. If one was going to visit the Paris collections and travel by train, the most sensible way was to take the Night Ferry from Victoria by way of Dunkirk which left London after dinner in the evening and arrived at the Gare du Nord the following morning at 8.42, in good time to allow one to start work, always providing that there was not a storm or fog in the Channel, in which case it was not unknown to wake up the following morning tucked up in one's wagon-lit in Dover Harbour.

If a band of us was going to Florence, for example, then the cost was astronomical. Looking as unnaturally elegant as store buyers of both sexes contrive to do, we would board the Golden Arrow on a Sunday morning and, as the Pullman cars trundled over the bridge which spanned the Brixton Road, we would wave a symbolic goodbye to what we called 'the Bon', the Bon Marché, one of the older Partnership stores which stood in it, while we waited for a steward to bring on the Bloody Marys.

In Paris, unless the train had been delayed by rough weather in the Channel, there would usually be time for drinks at the Ritz before boarding the night express from the Gare de Lyon on which a whole block of sleepers had been booked for us in a wagon-lits. Other acquaintances would sometimes join the train in Paris, among them Norman Parkinson, tall, thin, moustached and immensely elegant, who could be seen striding up the platform, preceded by a porter pushing a trolley loaded with great boxes of photographic equipment, on his way to record the Italian collections. And on one occasion there was a pair of fabulous twins who worked for *American Vogue*, complete with ladies' maids and Vuitton trunks. Because of this, dinner on the train was enormous fun, and no one went to bed until the small hours of the morning by which time, in winter, the train would be through the Alps and beyond Turin, ploughing its way through the fog and slush of the Lombard Plain. Then, around 7 a.m., beyond Bologna, where it was still dark in winter, it burrowed beneath the Apennines and emerged in Tuscany, in the meadows beyond Prato, where the sun would be coming up like a huge blood orange over what was then still, at least in part, an exquisite, generally fogless and often snowless, man-made landscape.

In July 1963, after sitting through the showings of twenty-six different boutiques in one day in the Palazzo Pitti in Florence, I was left with a dog-eared notebook filled with such enigmatic observations as 'sandals with chain balls round ankles', against which I had forgotten to record the name of whoever showed them. Although I did not know it then, this was the last time I would visit the fashion collections in Florence, Rome or anywhere else, as a buyer. This was the summer when a large, exotic insect flew across the vast white room in the Palazzo Pitti, biffed one of the buyers of I. Magnin of San Francisco on the nose and then fell dead at her feet, as if conscious of the enormity of its offence. 'You should put it in your memory book, dear,' said Russell Carpenter, I. Magnin's Director of Fashion Buying, who bore some resemblance to Noël Coward and attended each session of the fashion showings in a different suit of clothes.

That same afternoon, while sitting in the Sala Bianca, I received a cable from New York which read, 'Can you come New York as Associate Editor, Holiday Magazine? Reply soonest. Sions.' Sions was the man who had given me hell when the upper case button of my typewriter went on the bum. Nevertheless, I was tempted. Time seemed to be running out at the Partnership. The memoranda were showering in on me thick and fast from those twin scourges, the Director of Trading (Department Stores) and the Director of Buying, Fashion.

I decided to test American opinion on the spot; after all the Palazzo Pitti was crammed with it. I asked Russell Carpenter what he would do. I asked Hannah Troy, who always sat in the front row so that she could catch a strand of some rare fabric from a passing dress under her fingernail to take back to New York for *her* memory book, and who looked like an aged but affluent fortune teller on Blackpool pier. I asked Mrs Guggenheim Henry of Wanamakers, Philadelphia where the publishers of *Holiday* hung out. I asked Mrs de Mille of the

Associated Dry Goods Association and a number of others whose names I have forgotten. One and all – after ascertaining my salary with the Partnership – advised me to take off without further ado for New York and not bother to go back to London at all.

I arrived back in London from Florence at 6.30 a.m. on a nasty grey morning in early August to find a letter from the Chairman, giving me the sack. I had hoped to get my resignation in first, as I had more or less decided to accept the *Holiday* offer. I was now an ex-Partner, one of a distinguished body which included Audrey Hepburn, Mr Roy Jenkins and Mr George (now Lord George) Brown who had been in the Fur Department in Oxford Street. In the end I never went to *Holiday*. Instead, Wanda and I went to India and among other things travelled twelve hundred miles down the Ganges by boat, which took us three months.

23

Down the Drain
(1963)

1963 was a busy year. It was the year I went down the London sewers, while still a fashion buyer, the year I got the sack from the John Lewis Partnership, the year I went to America for the first time, the year I went to stay in a lighthouse on a rock in the Scillies, the year we went to India and floated down the Ganges.

I had been fascinated by drains on a grand scale ever since seeing Orson Welles clattering up and down brightly-lit Viennese sewers in hand-made shoes, hotly pursued by policemen in special sewer-going suits, a fascination that not even a visit to the well-scrubbed section of the Parisian *égouts* under the Place Vendôme, which is open to the public and deadly dull, could dispel.

I soon found that one of the principal pleasures to be derived from craving permission to go down the London sewers was that the authorities did not want to give it. The London County Council made this clear to me in a surprising hand-out which read as follows:

> Secretaries of religious, cultural, educational, political and social societies wishing to include a visit for their members to one of the Council's pumping stations or Northern Outfall Works or to have an illustrated talk on the main drainage service should write to the Chief Engineer. Applications to visit sewers cannot be entertained.

After conducting a protracted correspondence with the chief engineer, in the course of which I was asked *why* I wanted to visit the sewers, and to which I replied, 'Because they are there,' I eventually received a letter which ended, '. . . happy to assist you, subject to the Committee's approval, providing that your presence in the sewers does not increase the danger to personnel already working there.'

This letter, with its dark hint that my presence in the sewers might produce some sort of catalysis (something similar was suggested by the Elder Brethren of Trinity House when, later in the year, I was trying to get permission to stay in an offshore lighthouse), seemed more offensive on a first reading than it was intended to be.

Some time in February 1963 I took a day off from buying model gowns retailing at ten guineas upwards, and set off for my first sewer, King's Scholars Pond, better known as the Tyburn River. As a river the Tyburn rose near Marble Arch and it could still be heard, as I had discovered, gurgling away merrily beneath a twenty-four-inch manhole outside No. 3 Shepherd Market in Mayfair, the village in which, while still at St Paul's, I had had my first experience of the joys of commercial sex one afternoon after school. From Shepherd Market it ran under the dip in Piccadilly (which was really the valley of the Tyburn), and which was such an obstacle to the gun-carriage with the body of King George V on it when it was being hauled to Paddington in 1936, then under Green Park and Buckingham Palace and Pimlico, finally finding its way into the Thames by way of a disused dock, in which barges used to discharge cargo, behind the Sewer Depot in Grosvenor Road, near Vauxhall Bridge.

This last part was the only stretch of this unhallowed stream that was still open to the sky. And a good thing, too, I thought as Mr B, an assistant inspector of a London sewage area which extended from Flask Walk in Hampstead to Brondesbury and Tufnell Park, then to Blackfriars and from Blackfriars to Chelsea Bridge, led me along the muddy bottom of the dock to the underground part.

Mr B came from a family of sewermen. His father had spent thirty-eight years underground, while he himself had been at it for fifteen. He and his family lived in a charming Victorian cottage which overlooked the Thames opposite the Grosvenor Road depot, and the Tyburn ran right under it.

The bottom of the Tyburn was littered with some bizarre sorts of jetsam which included that morning a fine pair of unmounted antlers, a folio bible in the Welsh language, half a pram and an old bicycle – could they be from the Palace?

We entered the covered-in part. The water was not more than a foot deep, but the atmosphere was steamy. 'The steamiest of the lot is Piccadilly,' Mr B said, 'on account of the hotels and baths and washing-up going on all day and night.'

'What about The Palace?' I asked. 'Anything special about that?'

'Well, you can take it from me,' said Mr B, 'that what comes down hasn't got "By Appointment" on it.'

The truth was, as I discovered while I continued to make forays into them by day and night, that sewers are unpredictable, dangerous places into which, in addition to the more or less loathsome things they are constructed to receive, mad humanity pours all sorts of lethal stuff, all of it illegally. Even a short list of what found its way down the drains reads like a recipe for some twentieth-century witches' brew: acetylene; petrol, sometimes in large quantities; carbon dioxide, given off by hospitals and ice-cream factories; hydrogen cyanide from electroplating works, which has a nice smell of almonds, the faintest suspicion of which sent any gang of sewermen, who were known as flushers, straight up the ladders to the street; and as smells went, which I was able to inspire for myself in the Tyburn, most ghastly of all when mixed with untreated sewage: ordinary coal gas from leaky pipes. What North Sea gas smells like, if it smells of anything, when mixed with untreated sewage, I leave to a younger explorer to discover.

For these reasons, on initially entering a sewer, my first, almost overwhelming, impulse was to light a cigarette, or if I had had one about me, a Trichinopoly cheroot; but with the possibility of some or all of these things lurking in the atmosphere I would have stood a good chance, had I done so, of rising majestically through a manhole, rather like a Polaris emerging from a submarine. Some years previously a whole stretch of Kingsway had suddenly erupted without any human intervention whatever.

There was a narrow, slippery walk on either side of the underground section of the Tyburn and we had to take care, as we were not wearing waders. There were a lot of rats slithering purposefully up- and downstream along the brickwork, jostling one another like commuters in a subway. We were bound for a side entrance called the Keyshop in Tachbrook Street, Pimlico, one of the comparatively few places underground with a fancy name. There was one called the Corkscrew, at the river end of Northumberland Avenue, but only one sewer had a name on a board corresponding to the street above – Knightsbridge.

After pointing out one or two places where the brickwork had broken away and in which rats were nesting, Mr B seemed to grow as tired of the atmosphere as I already was. At first, I had been hoping to go north under Green Park and The Palace, but the fight had gone out of me. Instead we went up a narrow side entrance, and very unpleasant it was, to the foot of a vertical iron ladder.

'Blooming well stuck,' said Mr B, trying to force open the cover. 'Lend a hand.' A hand was not enough, but by pushing hard with heads and shoulders we suddenly opened the cast-iron lid and shot out on to the pavement at the junction of Moreton and Tachbrook Streets, SW1, to the surprise of a girl exercising a poodle.

More serious was the visit to the Fleet Sewer. Outside Blackfriars Bridge Station a van was waiting in the road with seven men in it, all dressed in waders, flat caps and donkey jackets, and drinking tea. There was a lot of washing water on the boil and bottles of disinfectant handy. There is a nasty complaint called Weil's Disease to which sewermen are prone, leptospiral jaundice, which is transmitted in the urine of rats, called by the flushers 'bunnies'. If it is not diagnosed correctly – it starts with a splitting headache and the symptoms of flu – it is usually fatal in twenty-four hours. It can be contracted through a scratch, which was why all the flushers I met washed themselves as religiously as any Hindu. Another disease is Miners Nystagmus, which affects the eyeballs.

This gang had come all the way from the depot so that I could go down the Fleet and they already had the cover up. They were nice without being garrulous. There was something about this job that made one want to open one's mouth as little as possible. They were glad, they said, that I was not an ambassador or some other sort of VIP, because they could only be taken down sewers that had been cleaned out specially for the occasion, and this meant a lot of work for nothing.

As soon as I had my waders on we went down a series of very slimy ladders into warm, steamy darkness, rather like a Turkish bath with something wrong with it. Again there was the awful smell that had made the Tyburn such a noxious place. Fulham Gasworks, I was told, was responsible. Whichever sewer I was in thereafter, all nasty smells were attributed to Fulham Gasworks and so far as I was concerned they could have closed the place down at any time they liked.

The Fleet at Blackfriars was a complex place. Great iron tubes spanned its upper levels, through which the station subways and the trains ran. Seen fitfully by the light of miners' lanterns and special lamps, it was like one of the prisons designed by Piranesi. Everywhere there were rusty iron gratings, long chains hung from the roof and the storm relief sewer was fitted with metal doors which, although they each weighed three tons, were held open horizontally when the flood water came roaring over the weirs, as if they were the tongues of paper envelopes. At the lowest level the Fleet itself raced riverwards at a good ten knots, too strong to stand up in on a dry day, down a tunnel more than fourteen feet high. When there was a storm, they told me, first of all an apocalyptic wind raged through the sewer and then the rain water came thundering down, filling it up to the brim, by which time they themselves would be brewing up in their van on the surface. The most surprising thing that had ever been found in the upper levels of the Fleet, the senior member of the gang, with thirty-seven years' service, told me, was an iron bedstead. No one knew how it got there, because you cannot get a bedstead down a household drain or even down a manhole without taking it to pieces, so someone must have taken it down in bits and reassembled it.

Best of all in my week in the sewers were the night shifts, after which I used to catch an early morning train to Wimbledon and undress in the garden before having a bath and breakfast and returning to Waterloo and the MG buying office in Oxford Street. One night I helped to flush out a four-foot sewer called the Opera, on the Embankment between Westminster Bridge and Charing Cross, which served Scotland Yard and other secret places, such as the Ministry of Defence. By eleven o'clock they had the manholes up in the side entrance in Richmond Terrace and a winch ready to haul up the iron buckets, which were called skips, to the surface. There were two top-men to work the winch and a lorry-man from a firm of contractors, who was waiting to drive the stuff away in the small hours of the morning and who was looking after the stove in the van. The men on the surface knew where the nearest telephone was in case there was need for an ambulance, the fire brigade, or smoke apparatus, or even all three at once. The nearest telephone was just over the railings inside Scotland Yard, where the policeman on duty watched our operations apprehensively, as if we might blow them to smithereens, something not beyond the bounds of possibility if we had wanted to. In addition it was the top-man's job to watch out for rain and get the gang up to the surface the moment it started. In the van there was a lifeline, lifting harness and first-aid gear.

Before the other five men had gone down, the ganger had lowered a Spiralarm gas detector lamp and a piece of lead-acetate paper in a wire cage into the sewer. If the light had gone out or the red warning light had gone on, or if the paper had changed colour, then it would have been dangerous to go down. None of these things happened, so the first man down took the Spiralarm with him and hung it in the sewer for all to see.

Down in the Opera we were bent double like convicts in a Siberian mine. I was wearing a deep wading suit which came up to my chest. 'Your lucky day, mate,' they said. 'Down the East End you'd be wearing a diver's suit.'

The sewage was nice and warm, just right after the freezing air on deck. It was up to my middle and the sludge we had come to shift reached my knees. The place on my face where I had cut myself shaving that morning was throbbing in a curious way and I wondered whether I was going to get Weil's Disease. I certainly had a splitting headache.

Everything was moving very slowly in the Opera, sewage-wise. This was because a band of loonies, masquerading as workmen extending the platform at Westminster Underground Station, had poured an enormous quantity of a special filling called Bensonite into the sewers and gummed up the Opera. The

London County Council was sending the bill for the flushing to the building contractors, but as one flusher said: 'Doesn't make me feel any better, whoever pays for it.' The gang was bitter because although the work was being paid for, a lot of Bensonite was still coming down. But in spite of everything, this was a good gang. They worked fast with their sleeves rolled up, up to their elbows in sewage, shovelling the Bensonite into the side entrance where the skips were. When it had dried off a bit they filled a skip and it was hauled to the surface. Apart from shouts of 'Orl right!' and 'Steady, mate!' they kept their mouths shut, like the men at Blackfriars. A night's work, from ten-thirty until seven the following morning, is four cubic yards of grit, which is a lot of grit to shift in a confined space. I tried to photograph them at work but, even with the fastest available film, without a flash it was like trying to photograph a band of spiritualists.

Later, when they had finished and were washed up and drinking tea in the van, they all said how much more healthy it was working in the sewers than being in the open air, especially with the weather being so nasty. By this time I was in favour of more material rewards, such as grace-and-favour residences in perpetuity, for the flushers, who got £12.15 ($34) for a forty-four-hour week.

I went to the Northern Outfall at Barking, where everything north of the river came out. Not even the most impassioned sewage engineer could say that it was a gay place. Admittedly, the landscaping had been done by the Parks Department, but even Battey Langley might have been up against it if he had to work on a sewage outfall in what had been a marsh in the Thames Estuary, itself set in a nightmare landscape peppered with electric pylons; and with Beckton Gasworks on the right and, somewhere on the other side of it but invisible, the river, with invisible ships hooting their way up- and downstream.

The route to the Outfall was by an interminable processional way devoid of trees, which reminded me of pictures I had seen of parts of New Delhi, except that a blizzard was raging. Apart from a man bent over the handlebars of a bicycle saying, 'Bastard, bastard, bastard!', as he ground along into the teeth of it, there was not a soul in sight. Perhaps he was the secretary of a religious, cultural, educational, political or social society on his way to fix up a conducted tour for the members, or an illustrated talk on the main drainage.

Inside the embankment on which the road was built were five horseshoe-shaped sewers which delivered everything produced by three million people in an area of a hundred and twelve square miles of London north of the Thames. It was an experience that I was not particularly keen to repeat, to stand over the detritus pits, surrounded by drag-line scrapers, conveyor belts, hoppers and dumpers – all the tools of this melancholy trade mercifully worked here by remote control – with my mouth clamped grimly shut like a captain on the bridge of a destroyer closing with the enemy, and to think what was passing under my feet: something that looked like mulligatawny soup with dead rats floating in it.

The detritus pits caught the ten thousand tons of road grit not removed by the flushers from the sewers in the course of a year. Eventually it was dragged from the pits and used for what was called 'reclamation', whatever that meant. On a dry day, and this was a dry day, 'on account of everything bar sewage being frozen solid', as someone vividly put it, one hundred and seventy-eight million gallons of the stuff flowed out of the five horseshoe tubes. On a nice wet day it rose sharply to two hundred and eighty million gallons. Between eight and nine in the morning, with everything in North London going full blast, a hundred and ten million gallons went into the main sewers in that one hour. It took seven hours to arrive at the Outfall from the furthest points west, out towards Kensal Rise and Shepherd's Bush. If it was high water for sewage at Kilburn at nine in the morning, it was high water at Beckton between three and four in the afternoon. Dead low water in London was around two in the morning, apart from the odd man from Fleet Street having a bath and Chinese restaurants still washing up, so that, in theory, the quiet hour for the staff at Beckton was around nine o'clock.

Most awful was the screen house where things that were not road grit passed through screens and were disintegrated. For the screen house only some disjointed notes in a discoloured notebook have survived, in which the word 'Things' appears frequently:

Pipes blowing Things back into the sewage upstream that have already passed through the Medium Screens . . . Bubbling sounds . . . Smells like Hell! . . . Looks like Hell! . . . Barrow loads of Things that failed to pass through the Coarse Screens . . . mechanical Rakes slowly lifting Things on which jets of water play until they fall off . . . House-proud men painting everything in sight made of iron a dull red . . . Told that the nastiest job is clearing blocked pumps. See man doing this, or something similar – Could be inhumane alternative to capital punishment.

Outside, in the primary sedimentation tanks which looked shallow but were eleven feet deep, seventy per cent of the remaining solids were removed for further treatment. The liquid part was now chocolate-coloured. How this could be when the last time I looked it was mulligatawny was a mystery.

At this point about fifty per cent of the sewage, thirty per cent of it sludge and Lord knows what else, went straight into Barking Creek by a sort of stage door and then into the Thames. When the Outfall was originally built no allowance was made for any treatment of the sewage at all. Four large reservoirs were built which could hold six hours' flow until the river was on the ebb, when it was all released, with spectacular results. At the time I was there the sewage men aimed to return to the river fifty per cent of well-treated sewage, the colour of mushrooms, free of sludge and apparently bounding with oxygen. This seemed pretty old-fashioned to me until I heard an Australian expert, who was on a grand tour of European sewage, say – with his teeth chattering, for the wind that day was straight off the Urals – 'That's nothing, sport. We let the whole lot out at good old Bondi Beach, just as mother makes it.'

The thickened sludge from the primary sedimentation tanks went through two processes of primary and secondary digestion – an unhappy term for these processes, especially for me, recovering from an unseemly lunch of overcooked sausages and custard wodge in the canteen. The sludge produced enough methane to operate a large power-house. The hard-core was pumped at the rate of six thousand tons a day to the end of what must have been the coldest jetty in Britain where, through a big black pipe that looked like an elephant's trunk, it sank into the holds of four sludge ships and was taken twenty-seven miles out into the Thames Estuary to the Black Deep. There, at the south-western end, in the spoil grounds marked by four conical buoys, they let it go. I did not ask the crews whether they enjoyed their work or not. They sailed nine tides out of fourteen a week, in all weathers.*

*These vessels were later replaced by a modern fleet of fully automated vessels, in one of which I subsequently had the opportunity to sail. They were spotlessly clean, as were the crew, although I would have hesitated to accept that the decks, as one of them contested, were 'clean enough to eat your dinner off'.

24

A Princely Shoot
(1963)

Before going down the Ganges we were invited to go shooting by two Indian princes who were nephews of the Nizam of Hyderabad.

At seven-thirty in the morning the sun was shining almost horizontally through the tops of the teak trees; the jungle floor was still in shadow. We were in deep jungle south of the Godavari river in Andhra Pradesh, in what used to be the State of Hyderabad before the Nizam was shorn of his powers in 1948. This was different jungle from anything I had seen before. Its floor was flat and sandy, and the little water that still came down from the *nullahs* between the hills spilled on to it in places, making it damp and cool. Nearby there was a lake on which, at dawn, we had seen ducks in flight. Beyond it was a village inhabited by Gonds – aboriginals – by nature honest and truthful but now, influenced by plainsmen, drunken and indolent; the men squatting in the shade while their women, who were extremely personable, toiled in the fields in the forest clearings. Near this village a tiger had killed a water-buffalo three nights previously, not for the first time and it had now been declared a cattle-lifter. In the village there was a woman who had been seriously injured a fortnight previously by a bison. Here people were apprehensive of wild animals. In the fields at night watchmen sat in little *machans* (watch towers) to see that the wild pig and the bison did not damage the crops. They sat in them all night from sunset until dawn with a little fire in a pot for company, alternately groaning and striking some metal object in the hope of frightening the beasts away.

There were tracks of bison everywhere in the forest and somewhere, if one could find them, there were big herds including magnificent black bulls that were absolutely fearless but very difficult to get close to. There were signs of tiger and panther, too, in the dried-up river-beds, and there were the marks left by the sloth bear: deep holes excavated in the search for beetle grubs, shattered ant-hills and the nests of wild bees from which nourishment had been sucked.

Now, without seeing a single bison or a sloth bear, we had been trailing another tiger that was supposed to be a cattle-lifter, trying to anticipate what it would do next. We had set off from the camp that the Nawab had set up in the middle of the twenty-four-thousand acres he had rented for this *shikar*, and had followed the dried-up bed of a river which was thick with the prints of tiger: a large one going upstream, a smaller one down. We crossed a mountain on which the rock looked like chocolate. Late in the afternoon we heard the ghostly hootings of long-tailed langur monkeys high in the forest trees. 'Sign of tiger,' everyone said, including the Gonds.

The big tiger was attacked and eaten by wild dogs, the most terrifying of all living creatures to other wild animals. They tore it to pieces as it ran. The smaller tiger killed a water-buffalo and its calf. It ate part of the big buffalo and dragged the calf eight hundred yards down a watercourse and into a thick clump of thorn in the middle of an open expanse of tall grass. 'We're going in to get it,' Owly said, and as we went in the words from a hymn, dimly remembered up to now – 'In Death's Dark Vale' – came back to me clearly as being curiously appropriate. The tiger was not there. It was not more than a couple of hundred yards away but it was not in the thorn clump. I was pleased.

The Gonds built a *machan*, three feet square, in the top of a rotten old tree above the dead calf. I spent eighteen hours crouching in it without moving, waiting for the tiger which was still not more than two hundred yards away. Instead of coming for the calf it walked up to the river-bed and finished eating the large buffalo.

We had been invited by the Nawab Habeeb Jung of Paigah to go on a hunting expedition to see how the company formed by a consortium of princes, of which he was vice-president, made the arrangements for well-heeled tourists to live the princely life in the jungle, having had a princely time of it in and around Hyderabad, including an elaborate reconstruction of a seventeenth-century Mughal procession, a Durbar in a tent, and a Mughal banquet.

What we were doing now was rather different from anything the paying guests were likely to experience. Before leaving on the expedition I had expressed reluctance to shoot a tiger, or any other wild animal. I was told not to worry. If a tiger was shot it would be a man-eater or a cattle-lifter, and the hunt would take place on foot: there was no question of a *battue*, as the Nawab and his younger brother, the Nawab Naseer Jung (otherwise known as 'Owly'), found it more exciting to walk them up.

Now, immediately ahead of us was a large humpy black quadruped. 'Sloth bear,' Owly said. 'Very dangerous. If it sees us it'll probably attack. You'll have to hit it first and you must kill it clean. They're terrors when wounded.' The Gonds, who had been with us up to now, were already high up in one of the two climbable trees. The Begum, Habeeb's wife, and my wife, were both plaintively trying to climb the other one which was already occupied by a forest guard, who was busy telling them that there was no more room.

The bear was some thirty yards away, just disappearing behind a thorn bush. It showed no intention of attacking me, but I was now the central figure in a ritual and there was no way out but to do what I was told. Praying that I would not miss I knelt down and waited for it to appear on the other side. I fired the left barrel.

The recoil was something I was unprepared for. The rifle was a .465 calibre Holland & Holland 'India Royal' with twenty-four-inch barrels, specially designed for Indian princes with a predilection for stalking big game on foot instead of sitting on an elephant and having them driven towards them. It weighs only $10\frac{3}{4}$ lbs, compared with the .600 Express, which was as big as an anti-tank rifle, fired a 900-grain bullet, and weighed anything from $14\frac{1}{2}$ to $17\frac{1}{2}$ lbs. Although the India Royal had the lowest chamber pressure of any large-bore rifle, it also had tremendous hitting power. The velocity at a hundred yards using a 400-grain, soft-nosed bullet was 2300 feet per second. This was Owly's rifle. At that time it cost £800 ($2240). His other rifle was a single-barrelled .375 Magnum Holland & Holland, firing a 300-grain bullet with a flat trajectory and delivering a knock-out blow not much inferior to the .465. These rifles, Owly's pride and joy, although he had others, were probably the finest heavy- and medium-bore sporting rifles in the world. The only practice I had had was with the .375, shooting at bottles on a lonely forest road which, I was assured, was a perfectly safe place for the purpose, although I thought it dangerous for such an exercise; and I was confirmed in this belief when, in spite of range precautions, six men rode up on bicycles. Life is cheap in India.

Instantaneously with squeezing the trigger I saw the bear rise up in a cloud of dust and then fall, out of sight. The jungle, which up to now had been quiet as the grave, became bedlam. Clouds of green parrots rose screaming from the trees. I started running towards the place where the bear had fallen, tripped over a stump and fell in a ludicrous position seldom adopted by serious big-game hunters, with both feet in the air. I picked myself up, feeling foolish, and went on. Owly was ahead; my wife and Owly's wife, the Begum, were still up a tree.

The bear was lying on its side, like a great hairy hillock. Blood was oozing from a point somewhere below the shoulder. It looked very dead.

'Watch it,' Owly said. 'It may be shamming. And there may be another one.' As he said this the bear began to get up. I fired the right barrel and it fell palpably dead.

At the same moment there was a terrible screeching sound. It was a sound that resembled nothing that I had ever heard. I turned, to see another, larger, even hairier bear tearing towards me out of the bush on all fours. It had a white snout and long, curved white claws; its teeth were bared and saliva was dripping from its mouth. It had none of the pop-eyed false *bonhomie* of the normal bear. Head-on, it looked like a pig wearing a fur coat. It was very angry and the person it was angry with was me.

There was nowhere to run to, no tree to scale. Anyway, sloth bears think nothing of climbing forty feet up toddy palms at night and drinking the toddy from the pots until they've had enough, when they drop off like great hairy footballs. It was the conventional situation in which one always imagined hunters to be, having fired both barrels and with no time to reload. The thing was now twenty feet away, coming straight at me. I had two more bullets but they were in my shirt pocket.

I was very frightened because I knew I was going to be torn to pieces in about ten seconds. For a moment I considered hitting it on the head with the India Royal but it would have been like trying to stop an express train with a stick; besides, incredible as it now seems in retrospect, I experienced a feeling of prejudice against using a rifle which cost £800 for such a purpose. It was not mine and it might damage it. This may seem unbelievable, but it is what I felt.

When it was about twelve feet away, there was an enormous explosion close to my right ear and a sudden spurt of flame as if someone had just lit a blow lamp. The bear uttered a terrible sound and fell flat in its tracks. Owly had shot it at point-blank range using the .375 from the hip.

I was deeply moved. 'You saved my life, Owly,' I said.

'I know,' he said. 'Lucky it stopped him. Lucky I didn't miss. The .375's only single-barrelled. I only had one round.'

In spite of being told that the bears had attacked a Gond woman I did not believe it and I felt bad about having shot them. Later, when we were back in camp, sitting in copper hip-baths, I said to him, 'Look, Owly, if there's anything you'd really like . . .'

'Well, actually,' Owly said, 'there is one thing.'

What, I wondered, did Indian princes who were nephews of the Nizam of Hyderabad ask for when they have saved someone's life – another India Royal to make a matched pair for lefts and rights at bison, something operated by clockwork from Fabergé, a saddle from Hermès, a deep-frozen showgirl from Salford?

'What I would really like,' Owly said, wistfully, 'is *Dog World*, the Christmas bumper issue. I can't get it here because of the currency regulations.'

25

Lonely Islands
(1964)

On our return to England early in 1964 we faced the now familiar problems of keeping alive while I wrote a book, in this instance *Slowly Down the Ganges*. I was therefore more than happy when Peter Crookston, then of the *Sunday Times*, proposed that I should visit some of the more lonely islands off the coast of Britain and record a way of life that was already in eclipse and which is now, largely, no more. Visiting lonely islands turned out to be an expensive operation, necessitating the hiring of vessels, some of them of quite large tonnage, from their sometimes venal owners, and on more than one occasion Crookston complained plaintively to me and to John Bulmer, a well-known photographer who accompanied me and whose ideas of travel expenses, as a member of Magnum (a well-known photographic agency), were rather grandiose, that we were spending money more quickly than the *Sunday Times* could acquire it.

> Winter is the worst season of the year, especially in the Western Isles where surrounded by the sea we are face to face with the worst elements of nature. Here we feel the Winter very lonely and dreary when nothing remains of the beauties of time past. Flowers have faded away and birds migrated to warmer climates . . . At times we do not receive mails for ten or fourteen days as no boats will venture across the windswept Sound . . . Sometimes we go searching for driftwood along the shore and very often we get a great deal. We boys do not feel the night very long at all. We play hide and seek round the houses, or play cards or read books and these all help to pass the time. In Winter we see plenty of geese and sometimes swans on the wing which is a sure sign of snow. But soon Winter will be over

and then we are into Spring with its long days and fine weather. Then the boats put out to sea and it is with unpleasant memory we think of Winter.

I found this essay on mildewed paper underneath a heap of thatching grass in an abandoned schoolhouse on the small island of Taransay, off the west coast of Harris in the Outer Hebrides. It was written in the early 1930s by a small boy who lived on the island. His laborious pothooks – heavily annotated by the teacher in red ink: 'Careless! Cramped! You must improve!' – tell us so much about the splendours and difficulties of island life that, reading them in 1964, with the population of the island down to five, he seemed a prophet of the inevitable.

By the 1960s the communities in the smaller islands of Britain were dying root and branch. Soon, unless the process of depopulation was halted, the only inhabitants would be sheep, Atlantic seals, birds and lighthouse keepers; the only visitors, shepherds and naturalists. Perhaps the naturalists might be pleased, but when the last islander left a way of life unique in western Europe would have gone for ever.

By the early 1960s, in the entire one-hundred-and-thirty-mile sweep of the Outer Hebrides, from the Butt of Lewis to Barra Head in the south, there were only six offshore islands that were still occupied by people who made some kind of living from the sea or the soil. In Orkney and Shetland the populations of the smaller islands had declined steadily over the previous thirty years. In 1951 the population of Foula, twenty-seven miles west of Scalloway in the Shetlands, the northernmost island in Britain, with its amazing cliffs twelve hundred feet high, was seventy-three. In 1961 it was fifty-four. Fara in Scapa Flow was down to four.

Almost all were difficult to reach. The instructions for getting to Taransay were like something from a comic novel:

Tarbert to Horgabust – twelve miles. Taxi. Try to hire the Rev. Macdonald's rowing boat. If he won't lend it light bonfire on headland and hope that the Campbells will pick you up.

The reality was even worse. The rain came down in sheets and you would have needed an incendiary bomb to start a fire anywhere, let alone on a headland. The wind blew Force 7 from south and south-west, and the Sound of Taransay was as impassable as the Strait of Magellan. It took five days to get to the island and then it was only possible in a large fishing boat from Leverburgh in Harris, fourteen miles away.

The owner took the boat in as close as he could and two men came off from the landing place in a rowing-boat. At first I thought they had come to take us ashore but they lay a few feet from us flourishing their oars. The elder one had a splendid beaky nose, rather like a puffin's. He was wearing a postman's uniform. He was the postman, not only to himself but also for the other four inhabitants. His son, who was in his middle twenties, was a larger edition of the older man. They looked hard men.

'You're not coming ashore.'

'I'm a writer.' It seemed useless to dissemble.

'You're a journalist. We had a journalist. Ach, the boggert! We gave him tea. He wrote about the tea.'

'I won't write about the tea. You can see what I write if you want to.'

Suddenly they capitulated and, laughing, let us get into the boat. The settlement was on the east side, facing the Sound. It was called Paible. Of the sixteen families who lived there before the war only two remained, four men and one woman. The last family left in 1942 but the drift away began in the early 1920s, at the time of Lord Leverhulme's grandiose schemes for the development of Harris as a centre for herring fishing.

There were only three houses on the entire island. There was the schoolhouse by the landing place, in which I found the essay written by the small boy; the school was closed in 1935, and the building was now used by the Macraes, the men who had come out to fetch us, as a store for lobster pots and gear. There was a grey and white two-storeyed house, alone in a great meadow, in which Roderick Campbell who, until recently, had been the owner of Taransay, lived with his younger brother Angus. And there was a magnificent 'black' house, next to the schoolhouse, in which Ewan Macrae, their brother-in-law, lived (he was married to the Campbells' sister) with his wife and son.

214

The black house – they were called 'black' because, before stoves were used, the interiors were stained with the smoke of the burning peat – was one of the most simple and beautiful dwellings ever devised, at least from the outside. Because there was no lime on the island the walls were built of dry-stone, with the spaces between filled with earth to keep the wind out. These walls were anything up to six feet thick, continuous walls with no right angles; everything was rounded, including the roof. It was a perfect shedder of wind. The roof beams were made from driftwood – there were no trees on the island – and the roof was thatched with bent grass, kept in place by a net weighted with stones.

The Macraes lived by lobster fishing; they also looked after the sheep and cattle for the new owner (Roderick Campbell had recently sold the island, together with all the stock, to John Mackay, who had a garage across the Sound at Horgabust where the Macraes went to fetch the post and stores). They grew barley, rye and cabbages.

The elder Macrae liked the island but was beginning to find the life a hard one. His wife, who was also an islander, felt the same. 'We've had enough,' they said. The son, Ewan, loved the island life but was ashamed of living in a black house. Like most islanders, he wanted to conform and live in a modern house with ornaments in the windows. He was unmarried. If his parents left, and there was no one to help him with the fishing and peat cutting, existence on the island would be almost impossible for him. The two Campbells, Roderick and Angus, lived a life of remarkable simplicity even for islanders. Neither of them had ever married and Angus, the younger, who was sixty-two, did the cooking on a magnificent old peat stove that was cracked from end to end and occasionally emitted clouds of smoke. 'We live on fish and tatties,' he said. Both brothers had a markedly quixotic appearance and were slightly disorganized domestically. They liked cuddy, blackfish. At one time the cuddy provided the light on Taransay and other such islands, as well as food: their livers were pressed and the oil was burned in an oil-lamp with the pith of a rush for the wick.

Roderick had recently fallen through the floor of the schoolhouse and torn a muscle in his knee. He could not get about as he used to, and now that he had sold his stock there was not much for him to do anyway. He was anxious about his knee. 'Do you think it will mend?' he asked. 'Do you think I was right to sell the island?' Both the brothers had voices as soft as the wind.

I told him I thought he had done right. 'You've none of the worry and if you want a bottle of whisky to keep the cold out you can have it.'

'Ah, well! Ah, well! I don't know. Perhaps I was wrong.'

The island had a strange, almost unearthly beauty. Behind the settlement the hills rose gently. Hidden amongst them were two small lochs with trout which Macrae, as a small boy, used to take with a worm for bait, like other small boys before him. In the burn that ran down to the sea there were millstones as big as cartwheels, with which the people used to grind the corn. All along the shores of the Sound were the remains of black houses, now roofless; it was difficult to believe that there were people still living who were born and brought up in them, for they could be as old as the Iron Age hut-circles on Dartmoor, as could the dark ribs of what were once potato beds, regularly spaced on the sides of the hills, like contour lines. On the beaches, the bones of whales, half-buried in the sand, whitened by the wind. 'The Norwegians brought them here some years back. They brought them here to whiten but they never came back for them and no one knows why.'

There were two cemeteries on the island. Once there were two chapels: one dedicated to St Taran, who crossed from Scotland to Ireland with three of his seven brothers on a stone made buoyant by prayer; the other to St Keith. Women were buried in St Taran's, men in St Keith's. Once, in error, a mixed burial took place and the next day the body was found above ground.

In the part of the island called Aird Vanish, to the south and west, there were great caves in which the sea sucked horribly at the saffron-coloured weed; there were natural arches and sheer cliffs with narrow ledges from which pale-grey, fishy-looking fulmars look down coldly. Here it was not difficult to imagine the irruption of the Kraken, the great sea monster of the northern seas.

Late at night, in the long twilight, the southern beaches of

Aird Vanish, with the spine of a schooner and its mast and spars cast high on the grass by what must have been the father and mother of storms, were like some place at the end of the world which, indeed, was what they were; curlews and oystercatchers uttered mournful cries, highland cattle, with horns like the handlebars of giant racing bicycles, loomed in the dusk. Two bodies were washed up here, one in the First World War, the other in the Second. They are buried together, high on the storm beach.

Fara, an island in Scapa Flow of only two hundred acres, was even more vulnerable to the threat of total depopulation than Taransay. Its working population was reduced to two: the owner, Mr Watters, and his wife, Ina. Mr Watters was sixty-two. He was born on Hoy, one of the larger islands in Scapa Flow, to the west. He left school when he was twelve and a half and worked as a herdsboy. 'After three months,' he told me, 'I was paid £1.80 [$5].' In the First World War he served in ammunition ships; in the Second, on a naval tender in Scapa Flow. He had no false ideas about the glamour of war. 'Everyone says that Churchill won it but where would he have been without all the men who went out of the Flow and never came back?'

His wife was older than he was but looked younger than her age. Neither one of them had had a day's illness in their lives. Mrs Watters was born and went to school on the island. 'I stayed on until I was thirteen to learn sewing and needlework,' she said. The school closed in 1946.

Before the war there had been twenty-five people on the island, but six of the men had never married. 'Bachelors have been the ruin of Fara,' Mrs Watters said. They had a son who worked in a fish-and-chip shop in Leeds, and a married daughter in Kirkwall, the capital of Orkney. They were a hard-working couple. Mr Watters fished for lobsters, but working the boat alone and lifting the pots was hard for a man of over sixty, however fit. He had three boats: a rowing-boat, another with an outboard engine, and a third, with a marine engine, that was slightly larger. Besides fishing for lobsters he also fished for cuddy off Switha, down towards the Pentland Firth. 'You can take them on a bare hook if it's a shiny one,' he said. They ate the cuddy themselves, smoked. They used to salt them, dry them in the wind, and then hang them up to smoke over the peat stove in the kitchen. Sometimes they had partans (crabs). Partans were virtually unsaleable but excellent eating.

They had three fields under cultivation: oats and turnips for fodder, cabbages and potatoes for themselves, three acres in all. There was not enough grass on the island – much of it had gone back to heather – but with fertilizers and more scientific farming the island could probably have supported three farms instead of one, and many more sheep. (The Watters used seaweed for manure.)

Mrs Watters looked after the lambs, but 1964 had been a bad year. Twenty lambs died and they lost a number of rams as well. She also helped her husband with the peat cutting. They could only cut peat in the spring and summer because frost spoilt it, and you needed a drying wind. The cutting was done with a long-handled peat knife and for anyone not used to it, it was a herculean labour – six cuts weighed half a hundredweight. They used to work at it for two and a half hours a day until they had cut enough to see them through the winter. It grew dark very early in winter, about half past three – Fara is in 58°51′ N – and it was not light until nine. In winter they sat by the peat stove in the kitchen and Mr Watters made lobster pots and his wife knitted socks. They had a wireless and an old gramophone. Mrs Watters was also the postmistress. Like Mr Macrae on Taransay, she got paid for the job. She was also in charge of the telephone-box, which stood in splendid isolation in the midst of nothingness, half a mile from the house. If you wanted to telephone them you had to write first. It would have cost them £14 ($40) to have it moved nearer the house and it did not seem worth it.

To get into the box you first of all had to remove a block of cement which was held in place by iron pegs. This was because in a southerly gale the wind forced the door open and sucked the telephone book out of the box. The winds were of fantastic strength. In 1953 a wind of over 125 miles an hour was recorded on the north side of Hoy; the only trees on Fara were a few plantations of battered elders – how these survived was a miracle.

Mrs Watters was very proud of her telephone-box, which was

certainly very smart. The Post Office paid her to paint it once a year. She liked people to use it because she was afraid that otherwise it might be taken away. It was their only link with the outside world in an emergency. I found the temptation to put money into it, in the same way as one put money into a collection-box for the repair of the fabric of a church, almost irresistible.

Mr Watters's only real relaxation was his weekly visit to the social club at the naval oiling station at Lyness on the other side of Gutter Sound. If anyone in Britain deserved a drink he did. He liked Highland Park malt whisky best, from the distillery in Kirkwall. 'Tastes of methylated,' was what he said of a well-known proprietary blend.

Mrs Watters did not go with him, partly because she did not like boats but really because she loved the island so much that she never wanted to leave it. She had not been ashore, even to Hoy, for more than a year. Were it not for this Mr Watters would probably have sold the island, and they would have been living in the small cottage, in the shadow of the oil tanks, that they had bought against the sad day when they would finally have to leave. 'It will break her heart when we have to go,' he said.

We crossed the Sound for the weekly outing in the middle-sized boat, the one with the outboard. The weather was filthy; it was blowing hard, wind against a strong tide; night was coming on. Mr Watters was a first-class helmsman; even so we bailed all the way across. There was a destroyer in the berth by the oil tanks and we had a wild night of it in the social club. The return journey was like the crossing of the Beresina or the Styx; fortunately the wind had moderated.

'I like a good laugh,' said Mr Watters when we were half way across, 'a good laugh and a good story. Nothing to worry about. I once took sixteen people and a coffin all the way to Flotta. That's where the cemetery is.'

Most of the remaining houses on Fara were empty shells, with grass growing on the roofs in which crows and pigeons nested. However, the one in which Mr Watters's father had lived was still in good condition. In it there was a great bed, like an open matchbox.

We crossed a headland on which there were many gulls' nests. On the way he killed a rabbit blinded by myxomatosis; he did it as if he disliked killing things. We came to a building that had once been the island shop, now reduced to rubble. 'There was a fire in this house. They never let it go out until the end. It burned for a hundred years.'

I told him about the American in Rome who, on being shown a candle that was said to have burned for a thousand years, promptly snuffed it with the words, 'Well, it's time somebody put it out.'

'I like that,' he said.

It was very clear weather. 'That's a bad sign,' he said. 'There's a pole on Binga Fea on Hoy – when you can see it, it means bad weather. Besides, I can feel it in my bones. My father always used to say he could feel bad weather in his bones, but when I was young I never believed him. Now I'm older I can feel it, too.

'That's where the German Fleet was scuttled in 1919,' he said, pointing northwards towards Cava. 'There and at the top of the Gutter. Seventy-one of them. It started about midday on June the twenty-second. By five o'clock they were all on the bottom. I reckon it was a put-up job. When they were doing the salvage work here there was a man called Cox, the firm was called Cox & Danks. It was a big job. The ones off Fara were in seventeen fathoms. They had to cut off the superstructures and the barbettes first. I used to hear Cox cussing and swearing in the middle of the Sound. "There's a stupid bugger born every minute and I reckon I'm one of them, taking on this bloody job."'

He had a great appreciation of natural things. 'I like to see the geese when they're on passage, on the way to Norway. It's a right lovely sight. I walk down to the shore so I can see it.' And of a duck carrying a weakling on its back, he declared it 'the most wonderful sight I ever saw'.

'When you've gone it will be terribly lonely for a bit,' he said. 'I like somebody to talk to and I like a good laugh.'

26

New York
(1965)

On my return from the lonely islands, David Astor offered me one of the most coveted posts in Fleet Street, that of Travel Editor of *The Observer*, and for the next nine years I led its readers a merry dance: down huge rivers; through feverish swamps and jungles; across deserts and savannahs; up pyramids, minarets and mountains, some of them holy, most of them inaccessible; into hitherto unvisited seraglios; down sinister blind alleys in search of Bulgarian tailors who made astonishing suits from black felt; and into previously unsung eating houses and cafés, not all of them salubrious. Nearer home I took them in search of grass-grown Roman roads and ridgeways, of follies and grottoes as well as tin and arsenic mines, canal tunnels and other even more exotic monuments of industrial archaeology. These were just a few of the places and things to which I contrived to conduct them, at least on paper.

What I liked about *The Observer* and still do, for up to the time of writing it is largely unchanged in spirit, was that unlike the Partnership, no one gave a damn what one did with one's spare time, and in all the years I worked for the paper no one ever told me that I was doing badly, just as no one ever told me that I was doing well.

What follows are some descriptions of some of the, to me, more memorable, not necessarily wildest journeys I made during what I can, without exaggeration, describe as some of the most diverting years of my life.

In the autumn of 1965 I went to New York. This, apart from the pleasurable but abortive visit to *Holiday Magazine* two years previously, was my first visit to the city, and my aim now in going there was to attempt to find out how the impoverished British could survive in it, faced with a strong dollar, and armed with nothing but their native wit and intelligence and a handful of old-fangled pounds.

Starting at the top and working my way downwards through the innumerable stratas which make up New York – and which are encountered long before you get down to the mud, sand and gravel, decomposed rock, schist, Inwood limestone and Fordham gneiss on which man-made New York is set up – took a long time, because everything was so different from anything I had known back home in Wimbledon, SW19. Even my favourite author, the American Richard Bissell, on whom up to now I had relied implicitly for information on the New York scene and how to comport oneself within it, was sometimes floored for an answer. The following passage from his book *Say Darling*, which describes how he came to write a musical called *The Pajama Game*, demonstrates this, as well as indicating that it was the sort of problem New Yorkers were having to face up to every day:

> In boarding a bus or any public conveyance should the gentleman assist his companion on first, or should he get on first? Suppose I and Miss Gloria Vanderbilt was boarding a bus to go have a Giant Idaho Potato at Toffenetti's and I fouled up the embarkation rites. She would tell Sinatra I was as square as a coffee table.

So that I would run out of steam before I ran out of money (I really was playing the game of doing the whole thing with a minimum of cash), and eschewing the bus and subway services for reasons that are immediately apparent to anyone who has ever been confronted by them, I carried out my investigations either on foot or by bike, starting on Fifth, Park and Madison Avenues. At that time you could still see, not locked away in huge automobiles but walking the streets of the city, something rare in Europe – where the waxy, exhumed, dandruffy look had spread like the Black Death – rich, beautiful girls dressed in garments by Norell, Chanel and Balenciaga bought from Bergdorf Goodman and Henry Bendell and similar emporia. In London, where the daughters of the rich were busy camouflaging themselves as members of the proletariat and taking elocution lessons in Birkenhead, they would have been the daughters of South American ambassadors. Here they were home-grown and could be seen not only in the streets but also up at the Plaza Hotel on Central Park – as I zoomed through the Palm Court around four p.m., too fast for the maître d'hôtel to sit me down to a muffin – taking tea with their mothers, some of whom were wearing the most fantastic hats. (One of the treats I was preparing for the penniless readers of my column was a free zoom through the Plaza.)

But as an un-rich, and now not so young Britisher, I had to get along without the rich girls and their even richer mothers, and what I found most fascinating about New York was the life of the streets in those huge, submerged areas in which I could see for myself the miraculous and painful process by which Chinese, Armenians, Transylvanian Jews, Negroes, Germans, Puerto Ricans, Italians and Gypsies (to name only a few of those who were undergoing it) became Americans while still remaining basically and stubbornly what they were. London had produced nothing like these great enclaves of immigrants, even at the height of the Russian pogroms, and not even the recent influxes of Jamaicans, Sikhs and Pakistanis had succeeded in producing anything like the huge concentrations of foreign groups that existed in New York. Even London's Chinatown down in the docks was no more. It had gone up in smoke during the blitz and had not yet risen again, as it was later to do south of Shaftesbury Avenue.

In New York, Chinatown flourished and Mott Street was filled with shops with what were, to a frivolous Englishman, names that somehow suggested the occupations of their various owners, such as the Lun Fat Co., Grocers; the Wing Woh Lung Co., Importers; and the Go Sing Coffee Shop. Here in Chinatown, unlike most of New York, the inhabitants really appeared to be enjoying themselves. And nearby, on East Broadway, where Grossman's Wine Shop announced that 'All our wines are made on the premises', Jewish territory impinged on Chinese.

The Jewish enclaves were more melancholy, presumably because the greatest humorists in America who were Jewish had fled from them. On Allen Street, where Jews from the Middle East made neckties for half the world – and you needed a sense of humour to be seen dead in some of their productions – I watched them playing cards after their day's work, sitting with their black hats on, on the upper floor of the Tirana Café.

On Hester Street they sold one another dill pickle and sturgeon by the chunk, and *kapchunkes* – whatever they were – were ninety-eight cents a pound. Not knowing about everything, such as not knowing about *kapchunkes*, was one of the pleasures of New York, letting the city wash over you like a mighty, highly discoloured sea.

On Essex Street there were shops, selling Jewish religious articles, that were so small and their stock so shop-soiled that one wondered how they could survive at all; and on the Sabbath, in what were, some of them, almost equally small synagogues, I listened to the readings in unison of the scriptures and watched the rhythmical rockings of the worshippers, many of whom emerged to attend these services from old red-brick tenements covered with fire escapes such as those on Broome Street, down towards the Williamsburg Bridge; there in the shop windows down at street level, men and women, some of them very old, sat on non-Sabbath days darning, weaving and 'stoteing', which were forms of invisible mending, just as, until recently, invisible menders sat doing similar work in a shop window in Piccadilly Circus.

Outside heaven itself, not even in Jerusalem was there any place in which so many Christian sects had come together, or one in which so many of their churches were massed, as in New York: the Reformed Secret Mission of Divine Research on Eighth Avenue; Macedonia God's Pentecostal on 111th Street; Hungarian Reformed on East 82nd Street; the Pentecostal Powerhouse Church of God in Christ on St Nicholas Avenue; the Chinese for Christ at 2274 Broadway; not to mention, because there is no space, this not being a classified directory, churches of Swedenborgians, Nazarenes, Abyssinian Baptists, and dozens and dozens of strangely named churches in Harlem and Spanish Harlem, such as the Iglesia el Encuentro con Diaz in a decayed area largely occupied by spare-part dealers and transmission repairers, down towards the Hudson off Broadway, above West 125th Street.

German territory, at least where many of them congregated

by day, was around Second and Third Avenues at East 86th Street. Shops here sold, as well as the beer steins, German and Swiss-German records (the latter making me thankful that my father had not executed a plan he had when I was young to send me to Switzerland to learn German), cuckoo clocks, *lederhosen* and female peasant garb with puffed sleeves. Where in New York did anyone, except possibly little children, wear *lederhosen*? In the privacy of some lovely schloss out in Queens? At the Berlin Bar, round the corner on Second Avenue, where they sold oversize *wursts*, I had a Münchner Weissbier with a slice of lemon in it, after which I felt like the Graf Zeppelin about to take off.

Two blocks away, around 84th Street, where the Hungarians lived, there were travel agencies with names such as Duna and Carpathia, and from this part of the city many parcels went out to places behind the Iron Curtain.

And up in the north was Harlem, as remote to white-skinned New Yorkers as Lhasa. 'Don't go to Harlem,' my friends said. 'Don't go to Harlem,' taxi-drivers said, as well as being actually reluctant to take me there as a fare-paying passenger. 'Go to Harlem,' counselled the thoroughly untrustworthy detectives in the Sixteenth Precinct House, which was far enough from Harlem for them to feel brave about it, pounding away with one finger on their hundred-year-old typewriters, filling in the charge sheets for the junkies in the small hours of the morning. 'Go to Harlem, but keep off the sidewalks, don't fool around in hallways [whatever that meant] and keep your eyes off the coloured girls.' This last piece of advice came from a member of the vice squad who looked as if he had been taking advantage of the facilities.

Confronted with these and other conflicting opinions – with those of the police, if one observed their prohibition about sidewalks, making such an expedition sound downright impossible – I decided to go to Harlem, in daylight and by bicycle, but some other year. However, the morning following my nocturnal visit to the Sixteenth Precinct House, I found myself riding my bicycle in Harlem by mistake, while travelling from the area at West 215th and 10th Streets where the sanitation department performed certain rituals which I had come to the conclusion would not interest my readers, even if they were penniless. It happened because I was ordered off the Franklin D. Roosevelt Drive, which was forbidden to cyclists, by cops who looked exactly like the ones who, two days previously, had suggested that I remove myself from the lower level of the George Washington Bridge on to which I had also strayed with my bicycle, and later from the upper level on which, as a result of their telling me to 'gederhelloutahere' without telling me how to do so, I subsequently found myself. It was thus that I found myself in Harlem, cycling along Seventh Avenue and St Nicholas Avenue, where I paid a hurried visit to the Pentecostal Powerhouse Church of God in Christ. Otherwise I kept my head well down over the dropped handlebars in an effort to avoid looking at the coloured girls, some of whom were making friendly gestures to the white cyclist, and who were wobbling like chocolate blancmanges in the process, obviously trying to lure him on to a sidewalk or into a hallway for a bit of fooling about. At the same time I also tried to look as if I were not the only paleface currently in Harlem, which in fact I was, except for those in cars with the windows up, who were all doing fifty miles an hour whenever they could, in an effort to get out of it.

New York was so different from London in so many ways that it would be tedious even to attempt to enumerate them, but to me the greatest difference between the inhabitants of the two cities was that New Yorkers seemed to suffer from time to time – but at indeterminate intervals – from a curious affliction that caused them to up sticks and swarm out of the parts of the city in which they had been living into other areas and finally, after a succession of such swarmings, out of it altogether. In London there never have been, and perhaps now never will be, such vast tribal, almost lemming-like movements, as those which in New York drove the rich and fashionable first up-town, then over to Riverside Drive on the Hudson, and then scuttling across to the East River, in a huge game of musical chairs with the poor; so that although still inhabited the blocks of apartments on Riverside Drive, for instance, once the wonder of the Western World with their Italianate wrought-iron lamps on either side of the doorways (at which no doormen in Ruritanian uniforms stood any more, as they still did outside similar blocks on Amsterdam Avenue) had no need of a sign proclaiming 'The Rich have passed this way.' There never was, perhaps, a city with as many ghosts of failed ventures and dreams split asunder as New York: ghosts of old department stores, between 13th and 23rd Streets, ghosts of old hotels, everywhere; ghosts of old newspapers, on Park Row; ghosts of Vaudeville, on and off Broadway.

27

A Walk on Broadway
(1965)

By day it may be Indian summer in New York but it is very cold up on the Brooklyn Bridge at 4 a.m. on an October morning with a salt-laden wind blowing up under the Verrazano Bridge from the Atlantic and into Upper New York Bay. It whips up the fires burning in the trash barges off Ellis Island which, now it is abandoned, is the loneliest place in all New York, with its long, empty corridors, full of old mattresses, in which the only sounds are those made by pigeons, a door creaking on its hinges and the bell buoys tolling interminably in the bay, rather like the abandoned harem at Topkapi. The wind blows strongly up on the bridge, and it plays on the miles of cable which support it, as though it was a giant aeolian harp. This is the hour when a friendly taximan – Mervyn Krmppf, I think his name was, give or take a few syllables – told me that the chances of being taken apart as an Englishman, or anyone else for that matter, begin to diminish. 'But you can't trust no one in this goddam town,' he said. Everyone gives you conflicting advice in New York about what you should or should not do. No one does this in England because no one knows what is safe and what is not. It is certainly pretty quiet up here on the bridge. Even the man who wrote 'I rape all the girls who cross this bridge' has had enough and gone home.

4.30 a.m. Down by the western approaches to the bridge, where grass-grown piers point long, decaying fingers into the East River, the Fish Market is going full blast. The fish arrive in big refrigerated trucks, each of which can haul 57,499 pounds. They stand outside Angelo's Fillet House and the Max Messing Co. Large hunks of tuna lie about on slabs and the scales are full of flatfish. There are big baskets of clams and oysters on the sidewalks. Braziers stoked with slats of fishy wood from old packing cases burn at the street corners. In the Paris Bar, which has a fine, carved back to it and lots of cut-glass mirrors, men in rubber boots knock back drinks and eat hot cereal and fillets of sole. It has a rugged, Victorian air about it which reminds me of Billingsgate Fish Market, its counterpart in London. Neither will last much longer; but compared with Billingsgate it is as muted as a Trappist monastery.

5 a.m. Reach Battery Park, where some poor, cold men are sitting on benches wrapped in newspapers, and take my first steps on Broadway, the longest street on earth and by no means the most attractive. It is also very draughty. The wind that streams into Upper Bay through the Verrazano Narrows cannons off the Statue of Liberty and then streams straight up Broadway. No. 1 Broadway is the office of US Lines. 'Don't miss the boat,' says a notice in the window, superfluously. From it and from the headquarters of other shipping companies high over Battery Place, later in the day the executives will look out from their suites, waiting for their ships to come home.

Between Nos. 1 and 2 Broadway is Bowling Green, a pocket-sized park watched over by a pigeon-spattered Abraham de Peyster who had a finger in almost every pie in the seventeenth century. Seen from it, Broadway, the longest street on this long, long island, is like the entrance to a labyrinth and the only living things at the entrance at this moment are Newby and one old tomcat.

By day at large, modern No. 2, the brokers at Edwards & Henly sit watching the prices flash on the board, dreaming of the day when the Dow-Jones average goes to 1000. On what lush beds are these brokers reposing now? A notice at the door says, 'For information ask Seymour Halper.' Should I call up Halper? There is no one else to ask except the old tomcat. Perera & Co. at No. 29 have 'Pre-pack foreign money . . . convenient and desirable,' and a window-load of Austrian ducats. Yet here the streets are not paved with gold. Instead they are dotted with little flattened discs of chewing gum. (In Siberia they are paved with the flattened caps of vodka bottles.)

5.15 a.m. on Wall Street. Even at this ghastly hour there are no parking places left. The place is ablaze with light. It looks as if it is being sacked. Men are staggering out of the buildings, not with bearer bonds but with waste paper for the garbage trucks;

others are going in with cases of Crystal Spring Water for the executive suites on the upper floors, and others in yellow crash helmets are digging up the road. I say 'Good morning' to these friendly groups, but I might just as well talk to that old cat.

At No. 23 Wall Street, the premises of the Morgan Guaranty Trust Co., even at 5.15 in the morning a sixteen-foot-high chandelier glows against the gold-coffered ceiling of the main banking room, and J. Pierpont Morgan, suitably embalmed in oils, glowers down from the south wall. Here, at noon on 24 October 1929, the bad day of bad days on which 12,894,650 shares changed hands, the heads of the five great banks met, and decided to pool their resources and support what was left of the market.

5.29 a.m. Back again on Broadway. It is dark, dark in Nassoit-Sulzberger and Co. Inc., Realtors; but the lights are on in the slender Trinity Church, parish church of the world's richest parish. They are on in the Bank of Tokyo and in the Gothic premises of Brooks Bros., where I could buy myself a button-down, polo-collar, long-staple cotton Oxford shirt for nine dollars if only they were open. Huge expanses of glass at the entrances to the office blocks are being cleaned by men using long-handled squeegees, and the Chock Full O'Nuts Sandwich House is being renutted for another day as I go past. Large men with bad feet, dressed in long, formless coats, lumber by in the opposite direction, yellow carts from the sanitation department are on the go now, and the world's most accurate public clock in the window of the American Telephone Company's building tells me that it is 5.31 and 20 seconds.

At 5.31 and 30 seconds I cross Fulton Street. Somewhere downhill to the right is the Fulton Fish Market where I was around 4.30 a.m. Fulton Street would have been a short cut if I had wanted to cheat. The lights are also on in St Paul's Church, where George Washington's pew is preserved. To the right is the elegant City Hall. Steam comes wreathing up through grilles in the road like incense from a host of subterranean temples, and from other grilles comes the rumblings of trains on the Broadway Subway which runs all night. No. 233 Broadway is the Woolworth Building, the greatest of the early skyscrapers. It looms into the sky complete with flying buttresses and gargoyles.

At No. 280, Modell's Shopping World sells firemen's boots, king-size Bromo-Seltzers, footballs, bar bells and Mongo Santa Maria-La Mamba belting Mr Watermelon Man on long play. Big food lorries bound for branches of Schrafft's and Horn & Hardart eateries zoom past, followed by three police cars with sirens hitting top C.

Enter the textile area north of Worth Street, dark region at this hour. Even darker are the streets which sink away eastwards to the Bowery; but coffee shops are lighting up now with their proprietors at the helm.

5.40 a.m. Cross Canal Street. Huge trucks are thundering eastward from the Holland Tunnel bound for the Manhattan Bridge loaded with anthracite, lumber, concrete pipes and syrup. Have quick coffee at Dave's Corner, 416 Broadway, twenty-four-hour luncheonette.

5.50 a.m. On the road again. Moon still high. The wind blows old newspapers ahead of me up this long street. A man is asleep in the doorway of Louis Bogopulsky, Drapery Fabrics. The terrible howlings of fire engines and, most terrible of all, that of the Special Emergency Truck, echo up the canyons from the east. Someone must be stuck in a manhole.

In the 580 street numbers now. Garment industry in sight. 'Urgent – operators wanted here for hem stitch and Merrow-Panties.' This is a long bit of Broadway.

6.10 a.m. Still dark to the east but a lighter dark. Meet woman going downhill in the opposite direction wearing decayed floral hat and gabbling to herself. Perhaps she is doing Broadway from its Canadian end for *Time-Life*. At No. 623 Kaufman's Surplus Arms has nice sign showing old-type bombs raining down. Who buys twenty-year-old bombs? Plod on past shops selling paper, nuts and bolts, green eyeshades, close-out lots of ballpoint pens (15 for 99 cents or 35p), surplus snow shoes, *knockwurst* (I thought it was *knackworst* – must be getting lightheaded), and nostalgic close-out lots of old election buttons 'Thos. E. Dewey, Kennedy and Adlai Stevenson for President.'

6.30 a.m. Reach East 10th Street. See first bus (empty) bound for South Ferry; also big, glossy girlie ad for Chemical Bank, 'When Her Needs are Financial, Her Reaction is Chemical.' The stinks master must have thought this one up. Here, in a restaurant with rustic decor, I could have a 'Jumbo-size hot Roumanian Pastrami Sandwich zestfully spiced' for 95 cents [34p], if it were only open. Papers are lying strung up at the news-stands now, all ready with the bad news. Greenwich Village to the west. See first bookshop at No. 828. Tempted by solid work, *Joseph Wood Krutch's Herbal*. US Hospital Supply Corp. have a nice selection of operating tables in the window. Reach Union Square where the sky is pearly overhead, honey-coloured over the East River, and deep mauve low down at the end of East 14th Street. Negroes are sweeping the square with huge brushes; old women are feeding the pigeons who have just got up; equally old men are going through the rubbish bins; and sewermen are coming to the surface – for a moment I look down into the rusty guts of New York before they put the lid on. Ahead, on Fifth Avenue, the Empire State Building is beginning to blush on its east face.

6.40 a.m. East 19th Street. The sound of ships' sirens comes booming up from the Hudson. Beaten-down-looking people, mostly old, are going to work now. The Empire State Building is like a Roman candle in the sun which is roaring up over the East River. Gift shop at East 21st Street sells 'Horrible Giant Monster Flies'. I am beginning to need a new pair of horrible giant monster feet.

7 a.m. Madison Square. Nasty wind at corner of West 23rd Street. Fine, period skyscraper. The Flatiron Building would look perfect with Harold Lloyd stuck on the face of it. Another has little crosses sprouting from the upper part, like a cemetery on the North Face of the Eiger.

7.15 a.m. Street lights are extinguished. Pass Knitgoods Workers Union Local No. 155 at 1155 Broadway, and shops selling outsize Puerto Rican underwear, terrible steins and gnomes, and Leda being given a crafty peck in square four. Around West 29th Street see first of the well-dressed: garment manufacturers in vicuña chesterfields, narrow-brimmed hats and knit ties getting in early to harry their designers. At No. 1255 they are offering twenty-five per cent off rocket missiles.

7.30 a.m. See more manufacturers at Gimbels' windows on 34th Street taking notes on the late after-dark dinner dresses as they go to work. Inside it is the 123rd anniversary sale and you can save twenty-three per cent on adjustable leg-loungers. There are also big reductions on 8.10 carat round diamonds, from $13,750 [£4910] down to $9895 [£3412]. At Herald Square, Macys, the greatest store on earth, stretches away forever westwards, two million twelve thousand square feet of it. It is jet-set week in the windows at the back on Herald Square, and they are featuring small women's suede suits in hunting tan piped in white suitable for a weekend's pheasant shooting with General Franco, at $225 [£80], but inside it is sale time and extra giant-size male mink hats are going for $59.99 [£21.40]. 'You are assured of fresh perfume because it is not returnable,' they say firmly.

At West 35th Street Broadway begins to tower. The lovely homes of Regal Knitwear and F. & P. Pflomm, Real Estate. At West 36th I pause to read inscription on façade of Greenwich Savings Bank:

Among the passers by, some go their heedless way to poverty. But you who wisely enter through these doors . . .

Well, whatever they say, there are no dwarfs staggering up the steps bent double under sackfuls of loot at this hour.

Pursue heedless, by now almost heel-less, way up Broadway. At West 37th see first policeman since leaving Battery Park, twirling a night-stick – all very well to be brave now it's broad daylight. Postcard shop branching into literature offers *Sex Life of a Transvestite*. Rails of beautiful garments being pushed up Broadway by Puerto Rican boys must make this the transvestite heaven.

At West 38th a Horn & Hardart automat offers special 'Dutch Treat': 'We make good apple pie. You pay for it.' Charming! At No. 1425 pass the Metropolitan Opera House, soon to be demolished, looking sad and neglected.

8 a.m. Times Square. The lights are out on the Allied Chemical Building where the news flashes go round and round, and where the manly sailor on the Camel ad who blows real smoke out of his mouth is having his lungs refilled. The lights are also out on the cinemas on 42nd Street. Although most of the human wreckage has floated away from the corner of Eighth

Avenue, there are still a few poor derelicts about, gravy on their lapels and baggy trousers, who look like Buster Keaton, and a lovely octaroon showgirl with two salukis on a white leash is high-stepping to a rehearsal. There is also a man smoking a cigar and pushing an empty pram.

The Broadway Bookshop is open – it only shut at 2 a.m. The Times Square Bowling Lane is open – it never shuts. The Garden Pharmacy, the only twenty-four-hour cheque cashing service in the world, is open and the twenty-four-hour news-stands are open – 'Rubirosa was fizzle in bed, Latin beauty says,' according to *Inside Review*. Toffenetti's at West 43rd which serves Spaghetti à la Toffenetti with meat balls ('a thousand yards of happiness', 'lucky was the day when Mrs Toffenetti walked down the archives of an ancient castle of the Count of Bologna'), is closed. The shop selling black-fringed home stripper kits – towels marked 'for after sex' and packets of 'phony vomit' – is closed and so is the shop selling stilettos – 'we despatch anywhere in the world'. The lights are out on Anthony Newley at the Schubert on West 44th and the Avalon Ballroom – 'For Folks over 28' – is as dead as mutton. They are out on Lindy's Restaurant, which has strawberries as big as snooker balls in the window. Presiding over this interesting mess is the statue of George M. Cohan. A pigeon is giving him a friction. 'Give my regards to Broadway,' he says. He said it a long time ago, and he looks as if he would like to take it back.

8.15 a.m. Breakfast. It is a beautiful day, the sun casts golden bars across Broadway which now resembles the chariot race scene in *Ben Hur*.

8.30 a.m. At West 53rd they have the National Pocket Billiards Championship, and next Sunday's Sermon at the Congregational Church at West 56th is 'The Banquet of Life' by the Reverend Dr Elfron Rees, by which time I shall be eating at home, all being well. People are streaming east to the business section across these golden intersections. Smart and sometimes beautiful girls are showing now in all shapes, colours and sizes, some of them as tall as skyscrapers in their simulated Male Mink Extra Giant Size Hats. A big cigar butt smoulders on the pavement, and this year's cars in showrooms make last year's models look like stone-age transportation.

A Walk on Broadway

9 a.m. Columbus Circus. Central Park South looks like a golden river, which is not surprising when you consider the rents. Westwards huge blocks are under construction, the high girders swarming with Red Indians who prefer this dizzy work to being down on the reservation. Blind people are tapping their way to the Jewish Guild for the Blind at No. 1880.

9.15 a.m. At 62nd Street make use of splendid (free) lavatory arrangements at the fourteen-acre Lincoln Centre, and see hanging bronze by Lippold composed of one hundred and ninety separate pieces, as well as Seymour Lipton's *Archangel*, which looks like a space vehicle which is, after all, what an Archangel really is.

March on. Blocks of apartments reach down towards Riverside Park and the Hudson, equipped with little awnings over the pavement and the doormen. North of 70th Street shoulders become squarer, hats more like pimples. Old Central European men doze on benches in the sun on tired little islands of grass, and Gitlitz at No. 2183 sells *wursts* as big as 75-mm shells. Here the buildings are neither high nor low, and they are built of sad-looking brick so that they seem to be in permanent mourning. The sidewalks are dustier, the shops are dingier, and drug stores begin to advertise 'Roach killer'. 'Did you know,' says a notice in a bookshop, 'that our West Side neighbourhood contains the country's greatest collection of works and artists,' and proceeds to back this statement up by an impressive display of books, all by local writers. On the corner the *National Enquirer* announces the latest news 'Behind Grace Kelly's Bedroom Door.' Kosher butcher announces that there is two days' Succoth, which is why there are so few people about. Some graffiti here: 'Don't blay with Blomberg,' one warns. 'Don't mes with Alpert,' says another, equally definitely.

10.30 a.m. Suddenly at No. 2898 a bright shop announces sheepskin coats at an alarming $145 [£51.80] – Columbia University is on the starboard bow. All at once, like in a musical about college life, the streets are filled with Columbians, and Prexys' which sells 'the hamburger with the College accent', is full of them too. The shops sell textbooks and Lautrec prints. I am jolly tired now, and the sound of the bells in the carillon at the Riverside Interdenominational Church makes me think I'm

235

dead and in an interdenominational heaven. Away to the left Grant's tomb rises mysteriously out of its own personal mist on what is an otherwise clear day.

From the heights of Columbia, Broadway topples downhill into the lost world of immigrant New York. At La Salle Street the Broadway Subway emerges from the ground and crashes across a girder viaduct sixty feet above the valley. On the steps of the decrepit brick tenements which face it, old men in carpet slippers sit watching even older Negroes collecting junk and loading it on to their licensed carts. Ahead Broadway stretches uphill; it seems to go on forever.

North of West 125th Street it enters a decayed area of garages, spare-part depots, transmission repairers and strange churches, such as the Iglesia el Encuentro con Diaz, which I have already visited. At 135th Street the Subway vanishes underground again and the shops get smaller and smaller. They sell 'Productos Tropicales', 'Chuletas' and 'Rubo Estofado', whatever that is. Papershops sell *Mundo Americano, El Imparcial* (New York edition), Jewish newspapers, and the Italo-American *Il Progresso*. There are lots of credit jewellers and shops selling plaster statues of the Virgen del Carmen. Out on the pavements the Puerto Ricans admire their children who are taking their first faltering steps on Broadway.

All this comes to a halt with the intrusion of a bit of old America at West 153rd. The huge cemetery where Audubon of the birds is buried, and in the next hollow is Audubon Terrace where children, who presumably cannot tell the time, are waiting for the one o'clock opening of the Museum of the American Indian.

11.45 a.m. At 169th Street Broadway for the first time for sixty-six blocks ceases to be straight and I am back on Jewish Broadway with terrible dark streets leading off from it. 'Eat Shunz's fish and live longer,' says a notice. Pass busy Leo Lichtblau, whose door proclaims 'Attorney at Law', 'Driving School' and 'Taxes', the 'Temple of Universal Light', and finally at 11.55 a.m. reach Luigi's Bar Restaurant, 4199 Broadway, at 179th Street, the last place before the George Washington Bridge and the bus terminal. The Discount Quality Cleaners on the other side of the road is No. 4198. But this is not the end of Broadway. It goes on, mile after mile, through Cuban Territory and then leaves Manhattan for the Bronx. The last identifiable stretch of Broadway is to be found in the town of Albany half way to Canada, but this is enough for one day. It really is a very long street.

28

Lawrence's Jordan
(1967)

Dist. E.M.		a.m.	a.m.					a.m.		
	Damascus ....dep	6 35	7 30 Mon, Wed, Sat.			Mecca		...		...
77	Deraa ....arr	11 10	12p23 ,, ,, ,,			Medina.......................dep	8 36 Tues, Thurs, Sat.			...
						Mahan..........................	6 p45 Wed, Fri, Sun.			...
...	Deraa ....dep	11a40	...			Aman	4 a21 Thur, Sat, Mon.			...
123	Samach (Tiberias)......	2 52	...			Deraa ....arr	8 a5 ,, ,, ,,			a.m.
164	El Fule (Nazareth) ...	4 42	...			Haifa.......................dep	...			6 0
177	Haifa..................arr	6 0	...			El Fule (Nazareth)........	...			7 17
...	Deraa.........................	...	12p55 Mon, Wed, Sat.			Samach (Tiberias)	...			9 13
138½	Aman	...	5 p37 ,, ,, ,,			Deraa............................arr	...			12 29
386	Mahan	...	3 a59 Tues, Thur, Sun.			Deraa ....dep	9 a50Thurs, Sat, Mon.			12p55
809½	Medina .... arr	...	3 p 9 Wed, Fri, Mon.			Damascus ....arr	3 p10 ,, ,, ,,			5 25
...	Mecca......................	...	In construction.							

Deraa ....dep	1 p30	...		Bosra Eski Cham .... dep	8 a15	...
Bosra Eski Cham ......arr	4 0	...		Deraa 326A .... arr	10 45	...

TIMETABLE OF THE HEJAZ RAILWAY, AUGUST 1914
BRADSHAW'S CONTINENTAL GUIDE

Screaming southwards down the Desert Highway from Amman at eighty miles an hour with a Copt at the wheel and the horn going full blast, scattering the local inhabitants, their flocks, herds and donkeys from the crown of the road on which they congregated with old-fashioned persistence, I wondered gloomily what would happen if we mowed them down or a tyre burst. Presumably, if we knocked anyone down in this part of the world, we would be slaughtered by the survivors, which I imagined would be the custom. If a tyre burst, at least we would be killed instantly.

Sometimes to the left of the road, sometimes crossing it to the right, and already undulating in the mirage, was the line of what had once been the Hejaz Railway, built in the 1900s by the Turkish Army on the orders of Sultan Abdul Hamid II, along the route of the pilgrim road, the Derb el Haj, from Damascus to Mecca. By the time the Sultan had been deposed in 1909, 809½ miles of this pious and strategically important work had been completed, linking Damascus to El Medina, the burial place of the prophet and, after Mecca, the most sacred city of the Muslims.

The telegraph poles along the line were the same sort that Lawrence's Beduin had so much enjoyed pulling down: cutting the wires, attaching the ends to the saddles of their riding camels and walking them away into the desert to the accompaniment of loud twanging noises. The culverts were the same cut-stone constructions that his experts had so much enjoyed blowing out with gun cotton. Even the engine of a train, when one finally appeared, pulling half a dozen assorted box-cars and flat-cars, throwing up columns of dense smoke, announcing its appearance long before it finally materialized, wheezing and groaning its way across the plateau, was the same sort of steam engine that they and the British technicians who instructed and accompanied them on their raids, had blown off the rails in dozens.

The line, and the road, traversed a wilderness that in places resembled a huge sheet of coarse sandpaper, in others milk chocolate that had melted and then set again. In spring it would have been green and filled with flowers; but this was August, the hottest month. Now the only vegetation left by the sun and the goats was as dry as cellulose at which, nevertheless, a few herds of camels nibbled lugubriously, but always on the move, in the hope of finding something better. They were guarded by young Beduin with rifles slung across their shoulders and as we roared past they held up shiny metal bowls, begging us to give them water. True to his Christian faith the Copt ignored them, unless ordered to stop, reluctant to relieve these infidels. We passed hills crimped like pastry, little encampments of black tents, phosphate mines, endless convoys of Mercedes-Benz lorries driving north from the Jordanian port of Aqaba, and what might once have been forts or *caravanserais* of the Mecca pilgrims, now nothing but heaps of stones.

When the great Arabian explorer, Charles Doughty, set off from Muzeyrib, the assembly point for the Mecca pilgrimage, forty miles south of Damascus – 'It was a Sunday, when this pilgrimage began, and holiday weather, the summer azure light was not all faded from the Syrian heaven; the 13th November 1876...' – the pilgrimage consisted of something in the region of six thousand persons, of whom more than half were serving men on foot, and ten thousand camels, mules, hackneys, asses and dromedaries, in a column two miles long, some hundred yards wide in the open places, headed by the Pasha of the Pilgrimage in a litter.

Occasionally, a few cranes would lift themselves out of this desiccated landscape and flap listlessly away into the greater wilderness beyond the road and the railway. Far to the right were the high hills above the Dead Sea. Once they had been covered with forests of oak, but when the Turks built the railway they cut most of them down to provide fuel for the engines. Afterwards the Beduin and the goats finished them off completely. Now more trees were being planted to replace them, or so it was said; but to restore these forests would be the work of generations.

We reached Maan – nine marches of the Mecca caravan south of Muzeyrib – now the end of the railway line, apart from a short spur that led off in the direction of Aqaba. Here we drank beer and nibbled sandwiches which were already curling in the heat. Maan was nothing but a collection of mud houses within an enclosing wall, clustered about a single minaret. Egyptian vultures preened themselves in a ghastly manner on the rubbish dump, contesting it with a number of brown-necked ravens. A flood had washed through the town the previous winter, and many houses and shops had been destroyed, together with a number of their occupants; but there was still an oasis of shade under the palm trees that had survived the deluge, and pomegranates, apricots and peaches grew there in season.

The station was some distance from the town. Opposite it there was a hotel, run by an Armenian, that had known better days a long time ago. I asked the stationmaster for a timetable as a memento but there had been no passenger trains for years, and now there was only one freight train a day, which came from Ras en Neqb, the railhead for Aqaba, on the spur.

In the marshalling yards there were long lines of ruined locomotives, painted in sun-bleached greens, yellows and reds. On their sides they bore brass plaques with the inscription 'Hejaz Railway' in English and in Arabic script. Some had the names of English makers on them, others were French, and one built by Sachsenmaschinenfabrik, Chemnitz in 1918 could only have been completed in time for the armistice.

For more than five hundred and twenty miles to the south of Maan the railway was wrecked, all the way to Medina; the water-towers, culverts and rolling-stock destroyed by the Arabs and the dedicated British demolition experts, of one of whom Lawrence wrote that if he ran out of explosive he would gnaw the rails with his teeth. Down there, locomotives were still lying on their sides by what was left of the track, just as they had done when the Turks surrendered. Fifty miles down the line from Maan the railway runs through fearful country to Mudawwara Station. Close to it, in September 1917, Lawrence pushed down the handle of an exploder which the navy had given him and, to his great surprise, blew up his first train. Now a British firm was said to be rebuilding the railway and eventually it was hoped to extend it to Mecca. All around the station at Maan there were piles of rails, points and sleepers, all new and waiting to be laid.

Beyond Ras en Neqb, nothing more than a heap of stones, the road climbed slowly to the escarpment of the Great Arabian Plateau. From it, one looked down two thousand feet to the Plain of Guweira. A strong wind was blowing across it, raising great clouds of dust and now, in the fearful heat of mid-afternoon, it resembled an enormous seething cauldron containing some pinkish preparation that was on the point of boiling over. Beyond it rose a jagged range of mountains and beyond them, according to the map, were the Gulf of Aqaba and Wadi Araba which, with the Dead Sea and the Jordan Valley, formed part of the Great Rift which extends from Syria to Mozambique. Beyond the Wadi and the Gulf were Sinai, the Suez Canal and Egypt.

To the left of the Plain of Guweira was the even bigger Plain of Hisma, another former inland sea from which the waters had departed, extending sixty miles to the Saudi Arabian border and far beyond. From its floor, on which a thin scrub grew, isolated rocks and larger islands of mountain rose a thousand feet or more, some with great dunes heaped against them by the wind so that they resembled petrified waves beating against islands in a petrified sea. Somewhere in this waste the Sherif Aid, riding with Lawrence for Wadi Rumm in September 1917, had gone blind in the glare of the sun.

The driver took us down from the pass in sharp swoops through a series of appalling bends, up which lorries from Aqaba were groaning to the railhead, and out across the plain. Columns of sand were whirling across the road and as they passed over it they left long licks on its surface which were in turn picked up, turned into more miniature columns and whirled away by the wind, which was like a blast from an open furnace. To the right of the road a dark drift of nomads on camels were reaching across the wind into the eye of the sun, all muffled to the eyes, heading towards the Sha'fat Ibn Jad, the jagged range we had seen from the plateau.

After about ten miles in this inferno we reached Guweira, a small ancient settlement where, in Trajan's time, the Romans had built a fort. It had also been a Turkish garrison in the time of the last Sultans and a camel-market, of the Howeitat Beduin, for longer, probably, than it had served either Turks or Romans. Now lorries outnumbered camels by a hundred to one. They were everywhere, parked in droves outside the tin shanties that served as shops and cafés by the roadside. It was as if a monster pull-up on some transcontinental American highway had been transported to the Arabian desert.

Beyond Guweira we left the road and followed a camel track that led, sixty miles to the south-east, to Mudawwara, once a station on the Hejaz Railway. This was the route that Lawrence had taken, riding out of Wadi Rumm with the mutually hostile tribes which he held together, for their first successful attack on a train – an attack which brought the Beduin so much loot that the whole force had been in danger of melting away and returning to its tents. This same track was also used by a battalion of the British Imperial Camel Corps when they attacked and took the station in August 1918, having crossed the desert from Egypt to Aqaba. We were only just in time to travel on this track and see it much as it had always been. Even then the Jordanians, with American aid, were making a motor-road out to the foot of Wadi Rumm, where a tourist rest-house was to be built.

Knowing this and seeing the Beduin swaying through the tamarisk on their camels that were more like plants on stalks than creatures, with the sultry red mountains to the right now in deep shadow, and the astonishing, isolated outcrops banked with driven sand looming up in the Plain of Hisma, I felt that I was lucky to be where I was before these changes took place. At the same time I felt cut off from any genuine sensation of desert travel, riding to Wadi Rumm in a truck, when I should have been racing towards it on camel-back; but it was too late for these nostalgic, juvenile longings; and anyway it was time to make camp.

Seen as we came into it in the early morning, with the sun shining across it, illuminating its western walls, the Wadi was a stupendous, unearthly sight. Twelve miles long and three thousand feet above the sea, it seemed not only large enough to contain the entire Arab Army of the Revolt, as Lawrence wrote, but also part of the Turkish one. On either side of its flat and slightly inclined floor, two miles across at its widest point, like the parting of the Dead Sea in some super-colossal film of the de Mille era, the opposing granite cliffs of Jebel Um 'Ishrin and Jebel Rumm rose two and a half thousand feet sheer in the air, capped with white sandstone that was like foam on their crests; while in the extreme distance, beyond the place where the cliffs ended, a single, detached mountain, Jebel Khazail, was also beginning to glow as the sun rose upon it.

In a remarkable passage in *The Seven Pillars of Wisdom*, Lawrence describes the temptations of Wadi Rumm and the mountain of Khazail:

> Landscapes in childhood's dream were so vast and silent. We looked backward through our memory for the prototype up which all men had walked between such walls towards an open square as that in front where this road seemed to end. Later, when we were often riding inland, my mind used to turn me from the direct road, to clear my senses by a night in Wadi Rumm and by the ride down its dawn-lit valley towards that glowing square which my timid anticipation never let me reach. I would say, 'Shall I ride on this time, beyond the

Khazail, and know it all?' But in truth I liked Rumm too much.

For six miles or so we ground on through the coarse sand until we came to a fort of the Desert Police. In the distance it had looked minute and now, standing in the lee of the cliffs of Jebel Rumm, it seemed a toy fort.

It was a square, archetypal desert fort, built of red sandstone and surrounded by barbed wire and shaded by a single casuarina tree. Outside the aristocratic camels used in the desert patrol browsed in a compound. The fort had two loopholed watch-towers and curious shafts on their outward walls which, in a full-sized fortress, would have been intended for pouring boiling oil or molten lead on the heads of an attacking force. Here, with molten lead and boiling oil hard to come by, and boiling water at a premium, they must have had some more mundane use.

The garrison consisted of a sergeant, a corporal and six troopers. They were dressed in long, almost ankle-length khaki tunics or coats with collars that buttoned to the neck, and red-and-white checked head-cloths (*mandil*) which were kept in place by a black cord (the *agal* or *maasub*). They were constantly fingering their head-cloths and never seemed to have them adjusted to their satisfaction. Around their waists they wore bright red sashes and they carried .38 Colt revolvers in holsters. The sentry on the gate was armed with a rifle. All of them were swathed in red leather bandoliers stuffed with ammunition, and in their belts they wore extremely sharp, silver-mounted daggers ornamented with semi-precious stones or glass, it was difficult to tell which.

Who had invented this extraordinary, romantic, dashing rig? Glubb Pasha, perhaps. Whoever it was, I half expected some British officer, fluent in Arabic, with a David Niven moustache, some good Second World War decorations, cherry-red trousers, desert boots, Viyella shirt and cavalry hat, to appear and start telling us about the warlike virtues of his 'chaps'. No one appeared. They seemed to be autonomous.

We were borne off by these walking ammunition dumps to drink coffee which they dispensed with effortless courtesy under an awning. They were a tough-looking lot, but like policemen everywhere when one comes on them with an aura of respectability (they had been forewarned of our arrival), they were extremely friendly.

They were all Beduin. 'No one else would stand it,' said the Copt, who was from the city; but by Beduin standards their life must have been heaven. Their kith and kin were close at hand in the long, low tents that were pitched just outside the wire. If domestic life became too much for them they could always retire into the fort until things improved. At the very worst they could volunteer for the patrol. It seemed a perfect arrangement.

Beyond the fort, in a sort of bay in the cliffs which here ran back in a deep rift, there was an encampment of a dozen or so tents of the Howeitat Beduin, who in this region numbered some eight hundred. The base of the cliffs was a chaos of rocks, some of them as big as a double-decker bus. From among these gargantuan heaps small trees sprouted. The air was filled with strange ululating noises, the cries of the herdsboys calling to the goats which infested the screes. The tent on the extreme right of the encampment was that of the sheikh, as was the custom. It was pitched close to the remains of a Nabatean temple, and while we were grubbing aimlessly among the ruins for potsherds, he sent word inviting us to visit him.

There, in the sheikh's tent, under the cool, porous goat-hair cloth which his women had woven, and with the side curtains cunningly adjusted against the sun, we squatted down, drawing our legs under us in the prescribed fashion, rolling about oafishly on the tribal rugs in our tight trousers, exposing our enormous Western feet.

More coffee was produced – here in Wadi Rumm I felt myself slowly drowning in it – poured from a tall pot that stood among the embers. It had a curved spout that made it look like an *art nouveau* bird. The coffee was delicious, better than that of the police, unsweetened and flavoured with cardamom.

The sheikh, Ayid Awad Zalabin, was a splendidly remote-looking man. It was difficult to guess his age. It could have been anything from fifty to seventy; but whatever it was, his children were both small and numerous and they stood in the sun outside the tent, pot-bellies showing under short shifts, sucking their thumbs and gazing in at these weird visitors until, at a word from him, they scattered and hid. From behind the curtains came vague, female noises.

After an exchange of courtesies, which took some time as they had to be translated backwards and forwards, I showed the sheikh some photographs of the Howeitat riding out to battle, taken at the time of the Arab Revolt; and another of Auda Abu Tayi (chief of the Howeitat at that time, better known to film-goers as Anthony Quinn) and of his son who, at the time it was taken, was twelve years old, and was still alive.

The sheikh expressed interest in them, but less than I imagined he would, and when I offered to give them to him he declined them. Even the spitting images of men and things dead and gone were of no consequence to him. He was only interested in them when expressed in words, as a story handed down, or else simply as a memory.

High above the tents, in the mouth of a rift in the mountain, there was a spring that supplied the fort with water through a long pipe which grew hotter and hotter as the sun climbed overhead. Above it on the rocks, a small band of men were rebuilding the Nabatean conduit which, in ancient times, had led down to a well near the temple. Whether they were doing this for practical reasons or as a tourist attraction was difficult to say.

The way to the spring, which was called el Shellala, wound upwards among the rocks past fig trees brutally hacked by the Beduin in their everlasting search for fuel, dwarf acacias and wild watermelon shrubs, the small fruit of which lay among the rocks as light as ping-pong balls, dried out and filled with black seeds that rattled when one shook them. On the path big black beetles pushed doggedly at bits of goat dung, stashing them away.

Until the pipe had been put in to connect the spring to the fort the water had spouted into a basin cut in the rock, but now that the pipe had been cemented in, it only emerged as a trickle. Yet it was still an enchanting place, a rare place in the wilderness, overhung with fern and deliciously cool in the shadow of the cliffs. It was here, while bathing after the ride to Wadi Rumm from Guweira, that Lawrence met the crazy old Beduin who cried out, 'The Love of God is from God; and of God; and towards God,' interrupting his bath and using the word love in a conjunction that Lawrence had believed Semites to be incapable of. Later, in the camp, he had tried to make him expound further but the old man uttered nothing but groans and broken words and afterwards he went away.

From the top of the ledge above the spring, ravines a thousand feet deep that were like narrow trenches, led away into the mysterious heart of Jebel Rumm. Here and there the rain of centuries had worn cup-like depressions in the rocks. Someone, probably a Howeitat herdsboy, had baited them with seeds and then balanced flat stones above them on sticks, so that a bird touching one of them would bring the stone crashing down and be trapped.

But now there was not a bird or beast to be seen. The place was utterly silent. And there was no echo. I shouted up into the ravines and against the walls of the cliffs and the sounds died instantly. It was like shouting into a blanket or in a storm, when the wind whips the words away and they are gone.

29

Treetops, East Africa
(1967)

When I went to East Africa for the first time in 1967 I took with me *Hints to Travellers*, published by the Royal Geographical Society in 1937, which was full of useful information, such as:

> In the Ruwenzori it was customary to give the porters a Cerebos salt tin full of *bulo* flour, besides salt and blankets. This corresponds to a native measure, a *kiraba*. Loads are made up to 45 lbs, rather less than the weights of loads carried on safari in the plains . . . Cutters who make a way in advance of the porters often go off at right angles to the direction indicated, either because the path is easier or for some reason known only to themselves.

Things had changed a bit in Africa since this was written. No one walked more than a few feet on the modern, motorized, packaged safaris; no African carried a minimal 45 lbs of your belongings on his noddle; no one gave anyone a *kiraba* of *bulo* any more; and instead of employing wood cutters to clear a way to the extraordinary Ruwenzori mountains you could contemplate them from the neighbourhood of the Mountains of the Moon Hotel at Fort Portal in Uganda – Telegram address 'Romance'. 'Cracking log fires. Nine-hole golf course. Pygmy village. Hot springs in which an egg is perfectly boiled by nature.' (Did you whisper 'four minutes please' as you dropped it in and hope for a piece of African magic?)

It is one's first encounter with wild animals, other than animals in cages, that really sticks in the mind, however banal the circumstances. In my case, in Africa, my first exposure to them took place after Sunday curry at the Outspan Hotel at Nyeri, a morning's drive north of Nairobi in the lands of the, until fairly recently, not all that friendly Kikuyu tribe. We piled into Land-Rovers – not really necessary for such a trip but they gave an atmosphere – and set off for Treetops, the lodge at the foot of the Aberdare Range. After crossing a six-foot-deep ditch dug to keep the elephant out of the surrounding farm land, we walked the few hundred yards to the lodge, accompanied by an ex-Indian Army colonel, armed with a rifle. There was a strong feral smell in the air.

'Keep together,' he said. 'There are elephant about.' The Japanese in the party, who were loaded with a multiplicity of cameras and lenses, had what was surely a dangerous tendency to lag behind and point 500-mm lenses at little flowers. Sure enough, there on the track were four large dollops of what looked like Old Auntie Mary's rich Dundee cake still steaming from the oven, the new-laid droppings of an elephant. Everyone was impressed and one of the Japanese photographed them in colour. The rest of us scuttled after the colonel.

The new Treetops, unlike the old one in which Princess Elizabeth became Queen and which was burned down in the time of the Mau Mau, was not really in a tree at all, although branches writhed unexpectedly in the corridors. It stood on piles above a large pool and salt lick, the edge of which was so trodden by animals that from the upper floors of the lodge it looked like an aerial view of Passchendaele.

Treetops was all right. Had it not been to our liking it would have been just too bad, because we were locked up in it until the next morning.

On the far side of the pool, warthogs and their young were zooming round in circles; a huge, rare, black, giant forest hog was looking at a battered tree as if deciding whether to demolish it or not; and up on the sun deck of the lodge there were baboons, with inflamed faces and even more alarming effects at their other ends, careering about, knocking over loaded Pentaxes, pinching the sandwiches laid out for afternoon tea, and disappearing rudely between ladies' legs. In addition, black-headed orioles in the Cape chestnut trees made fruity noises, and thousands of weaver birds were seething away in a bed of reeds in the middle of the pool. The noise was terrible. We were told to keep quiet so as not to frighten the animals. It seemed a superfluous warning, like telling children at a cocktail party to pipe down.

But there was nothing to the bedlam which broke loose when night fell, and the baboons and their young had departed, and the weaver birds had taken to their swinging nests in the reeds. There had been rain in the bamboo forests high in the Aberdares, and to escape it the animals had come down in force. At any one time throughout the night there were fifty elephant outside the lodge under the floodlights, all taking up trunk-loads of mud packed with health-giving mineral salts; black buffalo wallowed in it so deep that only the huge black handlebars that were their horns showed; rhinos wallowed less deeply – all were the uniform saffron colour of the mud. All species, including the various sorts of buck, observed a wary apartheid. As in the world of men there was a lot of confrontation and a lot of backing down – everything feared the buffalo. Only the rhino really faced up to one another, and when one of them slipped his opponent a length of allegedly aphrodisiac horn, it went lumbering off into the forest, groaning.

It is the noises they all made that I remember: gaseous noises; sounds like heavy furniture which has lost its castors being moved across a room; the sounds of the last water going down the plug hole; even more weird gurglings and the sound you produce when you blow into a funnel.

Treetops was perhaps the one hotel in the world where, if you could only get to sleep, you could share a double room (you had to whether you liked it or not) and snore to your heart's content; but sleep by night was impossible and a waste of valuable viewing time. When the swift African dawn came around six-thirty, the animals were still there, stuck in the mud, and I was hooked on East Africa.

30

Orient Express
(1969)

In the depths of the winter of 1969, winter being the best time to visit a great city, particularly Istanbul, and a hundred and thirty years having passed since Miss Julia Pardoe wrote her valedictory paragraphs, we set off to see for ourselves what had been going on in those parts.

We travelled by the Orient Express: 'we' because it is inconceivable that anyone should spend such a bundle of money to travel in a first-class wagon-lit alone, or for that matter with anyone with whom one was not on terms of the deepest intimacy.

Of all the couple of hundred named Continental expresses it is the Orient which has most kindled the public imagination. Perhaps it is the thought of beautiful women in bed thundering across Europe towards the capital of a tottering empire presided over by the half-mad, pistol-toting Sultan Abdul Hamid II, who had walled himself up in the Yildiz Park which he had constructed by entirely eliminating a large Muslim cemetery.

The inaugural run of the Orient Express took place in October 1883, when it set off from the Gare de L'Est with an extraordinary assortment of persons on board. Among them – a representative of the Sublime Porte; a marvellous but very hairy chef; M. Nagelmakers, the Belgian founder of the International Wagons-Lits Company; an author and various journalists including *The Times* correspondent; Henri Stefan Opper de Blowitz, 'under five feet high and more than five feet round the waist', who on his arrival in Constantinople scooped the world with the first recorded interview with Sultan Abdul Hamid II. At Strasbourg they were joined by two superb women, one of them the wife of the Austrian Minister for Railways. The whole party travelled in luxury which is inconceivable today.

Then, as in 1969, the Orient Express did not go to old Stamboul. After Bucharest it stopped on the shores of the Danube at Giugiu where the passengers crossed the river on a ferry to Rustchuk and then went on to Varna on the Black Sea by train in seven hours, from where it was fifteen hours to Constantinople bucketing in a steamer, a total of 81½ hours. This was thirty hours quicker than any previous service. The train had to follow this rather circuitous route because the Bulgarians would not allow it to run through their country. They were still bloody-minded about railways in 1969. Even then there was no restaurant car on any international expresses running through their country.

The Simplon-Orient route to Istanbul by way of Venice and Yugoslavia was opened only in 1919. In 1969, so far as the Orient Express was concerned, Bucharest was the end of the line.

For this reason, in order to make a sentimental pilgrimage on the Orient Express and at the same time get to Istanbul from Paris, it was necessary to change trains and stations at Vienna. Before catching it we spent two very agreeable nights in Paris at the Hôtel Vendôme, in the Place of the same name and with a splendid view of it, hemmed in by Second Empire furniture, and travelling up and down in a lift so small that it was like being in a matchbox lined with red plush – being in the Hôtel was like looking at a Ritz through the wrong end of a telescope, and looking at the bill through the right end of a burning glass. It cost F77.50 (£5.80 or $13.90) for a double room, but was worth every centime of someone else's money.

The Orient Express left the Gare de L'Est at 22.15. Before boarding it we had a highly ritualistic and excellent, if rather copious, dinner, at the Relais Gastronomique Paris-Est, on the first floor in the station, having broken through a checkpoint where the whole question of whether we should be allowed in at all was carefully deliberated. Just because you happen to be interested in train-spotting does not necessarily qualify you to beggar yourself by dining at the Relais Gastronomique Paris-Est, without previous vetting by the Direction. We got in because we were both wearing the kind of fur-lined great-coats that are mandatory for travelling in the Balkans in winter, and also for residing in Istanbul.

Le Diner
Pâte de Pistache
Langoustes Soufflées
Noisettes D'Agneau Charmereine
Mousse Glacée
Wine: Quart de Chaume and Chinon
Eau-de-Vie

No audible conversation, let alone laughter, allowed and no discount for leaving your Michelin on the table. The management had not read the bit beginning '*Entrez . . . votre guide a la main, vous montrerez ainsi qu'il vous conduit en confiance*', and if I had done so would have sent us to the station buffet.

After this memorable repast, which took hours and which could easily have qualified it at that time to be included in the *Guinness Book of Records* as the most expensive railway restaurant in the world,* we withdrew our real leather luggage lined with silver-topped crystal bottles and ivory-handled hair-brushes, which Wanda had bought in a sale because they all had the same initials as she had inscribed on them (we were really doing this in style), from the depository, pushed our luggage, there being no porters at this time of night, up the platform of this sad, sad station, with its memories of Verdun and long-dead *poilus*, and boarded the Orient Express. The exercise did us a world of good.

The next day a German dining-car, which seemed to run on velvet, was attached to the train, although the wagon-lits conductor was Austrian. In Vienna we had eight hours in which to buy luxuries in the way of food and drink for the next two days and dine again, an indifferent dinner at Sacher (where I have never dined well, although I always hope to), before catching the Balkan Express from the Sudbahnhof at 22.20. The wagon-lits on this train was a very gruesome affair, no mahogany, no rare inlaid woods or brasswork. They were made of what appeared to be tin that had been painted a sickly shade of blue.

*The most lushly decorative station restaurant in Paris is the Train Bleu at the Gare de Lyon, which is embellished with frescoes depicting the journey from Paris to the French Riviera.

The only good thing about the Balkan Express was the conductor, an archetypal wagons-lits man with a permanent six o'clock shadow and the conspiratorial air of a *valet de chambre* who has someone with uncertain tastes for a master, which wagons-lits conductors often have. He told us hair-raising stories about the route before the war, addressing us in English but calling us Monsieur and Madame. How, in 1929, the train was snowed up for seven days in the highlands of Eastern Thrace, surrounded by wolves and how they burned the furnishings *lit* by *lit* to keep warm (which was not true) and about the prostitutes you could telegraph down the line for and toss off the train in another country when you had had enough of them, just as long-distance lorry-drivers do now, nearer home (which was).

Zagreb, the following morning, was full of coal-burning locomotives belching smoke and steam and when the train left two members of the Yugoslav Secret Police travelled with it in order to impound the film which I had exposed on these engines. By an elementary piece of sleight-of-hand I was able to hand over an unexposed film and then embarrass them by demanding a receipt. Neither of them could write. Hereabouts, other fragments of trains were attached, of interest only to real train enthusiasts, and our part became the Istanbul Express.

The dining-car was decorated in Balkan Modern – it was as if Pola Negri was about to burst in on us triumphant in furs – but some hidden instrument played gloomy music interminably. The only currency that caused any animation was the Deutschmark. Outside, the weather was bitter and we saw great skeins of geese flying high, boys playing ice-hockey on ponds, using brooms for sticks, and villages whose single street was a frozen mass of mud. Outside Belgrade the train stopped at Topcider which had a station-house like a Habsburg hunting-lodge.

At the Bulgarian frontier the dining-car was taken away and the American diesel engine replaced by a huge steam locomotive, which panted up through the snow and over the Dragoman Pass in the darkness. In our *voiture* we were the only passengers now, ripe for a sticky end. At five the next morning the train crossed into Greece. In Macedonia, big Greek steam engines were standing in the sidings under the arc lights, belching smoke. It was bitterly cold. The place to get a taste of it was in that unclaimed territory over the couplings between the carriages, where the coal burning wagons-lits stoves couldn't reach. This was (and is) wild, dangerous country on the frontier with Turkey. There used to be bandits, now there were watch-towers and minefields and the line was heavily guarded by soldiers wearing heavy, hooded coats and sub-zero-type boots. Great drifts of snow and icicles hung over it. To the left there were frozen swamps and a big river jammed with ice.

With the sun rising downstream over Thrace we entered Turkey and at seven-thirty reached Edirne, or rather a station near it. There were beggars, many of them children, all along the line in the bitter wind, swathed in rags. We felt ashamed to be travelling in such comfort. Then we crossed into Greece again, and out of it at a place called Pythion, where a dining-car was attached. Nothing to do with the Wagons-Lits Company, it served an awful lunch. It was as if Turkish Railways were trying to destroy us but not quite hard enough to succeed. In the afternoon the train, now with two steam engines in front, passed through Çorlu, near which the Orient Express had been snowed up in 1929, and battled up a series of immense curves through the heavily fortified zone east of Cerkeskoy. Out with the telephoto lens and the notebook!

At 15.15 we arrived in Istanbul, 2053 miles, 63 hours and 38 minutes from the Gare de L'Est, which included eight hours in Vienna; but by the time we had secured our registered baggage from the customs authorities and secured a beat-up American automobile of the Steinberg era masquerading as a taxi, the sun was setting.

'Pera Palace Hotel,' I told the driver.

Out on Galata Bridge a strong wind was blowing down the Golden Horn, carrying with it the smell of tanneries and worse. The Asma Sultana's *arrhuba* does not stop at the Sweet Waters of Asia any more and there are no Greek maidens to dance the *romaika* on St George's Day; but looming in the twilight, like a cut-out against the western sky, was the fantastic, incomparable sky-line of old Stamboul; the vision, if ever there was one, of a battered paradise.

31

The Pera Palace Hotel
(1969)

We sat in the Bar Americain of the Pera Palace Hotel in Istanbul, recovering from our journey through the Balkans on variants of the Orient Express. The bar itself had a shiny brass rail round three sides to support the feet of those whose arches had given way under the strain of attempting to carry out to the letter Herr Baedeker's pre-1914 instructions on what should not be missed in the city, although I had never actually seen it used for this purpose (in fact I had never actually seen anyone sitting at it).

There were three bell-pushes by our table, and when I pressed the one marked 'Barman' and he appeared, I was tempted to ask him why no one ever sat at the bar, but I knew that his answer would have been because there are no stools. Instead I ordered a bottle of the good red wine called Buzbuq which we had first drunk from a hock bottle in Erzerum twelve years previously, although it would have been cheaper to have bought it at the shop on the corner opposite and drunk it in our room – a practice which is not fair in a hotel such as the Pera Palace if carried to excess, for reasons that will be revealed, but is commendable to all except the weak-minded in, for instance, a Hilton. The other two bell-pushes were marked 'Groom' and 'Garson', so that besides the Buzbuq I was also tempted to order up a horse and an omelette.

The Pera Palace was built in 1898 by the Wagons-Lits Company at about the time when the Orient Express began to run as a *train de luxe*, twice weekly from Vienna, to what was then Constantinople, by way of the Balkans, in seventy-three hours. Much less sumptuous than, for example, the Bristol in Vienna, where guests were so smothered in comfort that they find difficulty in assuming a vertical position, less exotic than the Grand Hotel Oloffson at Port-au-Prince, in which one's last hours on the premises are spent scheming how to leave it without tipping a horde of Negro servants already adequately provided for in the bill, it was until recently my favourite hotel, a melancholy oasis in the eerie labyrinth of Pera, although very few of its rooms now have views of the Golden Horn, other buildings having mushroomed up around it. In 1979 the Pera Palace was closed by a prolonged strike imposed by outside forces who picketed it. It was the antithesis of what most other hotels have already become in these years before the next breakdown when the only thing to stay in will be a concrete bunker reached by an armoured train. (It, too, I am sorry to report, has now gone what seems to be irredeemably 'off'.) The bedrooms, reached by a wooden lift with silver-painted gates, were a good twelve feet high, the walls were more or less soundproof, and large pieces of furniture, outmoded to the point of being newly modish, loomed over brass bedsteads with knobs on in which the sheets were changed daily – at least ours always were, because we asked for them to be. These great *armoires* were inset with enough mirror-glass to satisfy all but the most jaded tastes, and the chandeliers and candelabra were collectors' items, which is why the worst sort of guests sometimes made off with the irreplaceable glass shades.

They would have probably taken the taps, too, which were nickel-plated brass, made by Horcher of Paris, if they had had the tools to take them off – what a place for a plumbers' convention! And the baths which they fed were so commodious that, once inside, even the most abundant occupant disappeared from view. The bidets, built on a similar scale, did not engulf the user as completely.

When we first stayed at the hotel, on the way to Persia in 1956, the bath in the room we occupied had the unusual facility of filling itself by way of the waste pipe with water of a rather curious colour, without our having recourse to taps; and we watched this process enthralled, thinking it must be something to do with the currents in the Bosphorus, until we discovered that it only happened when the man in the next room pulled the plug out of his bath. The most eerie suite was the one occupied by Ataturk when he took time off from trying to hoick Turkey into the twentieth century, and was unchanged since he used it.

What was described in the brochure issued by the management as the Salle des Fêtes was the hub of the hotel. It rose through the first floor into a ceiling full of cupolas, and was embellished with brown marble, huge chandeliers of eastern inspiration and on the floor, in the middle of it, there was a great brass vessel for burning incense. All round the first floor beneath the cupolas, latticed windows, as in a harem, opened out on to little balconies, some of which offered a distant view of the reception desk.

One of the things that made the Pera Palace different from other hotels was the staff. If anyone retired or died he was immediately replaced by a moustached facsimile who remembered us from our last visit when he was not there. The female staff never retired and never died. The food, as it always had been, was pretty dull, if not terrible. In fact, when I once wrote extolling the virtues of the hotel, several readers wrote that they thought it was the worst hotel they had ever stayed in and that I must be in the pay of the management to write as I had done, which only goes to show that among travellers tastes differ.

Another thing that distinguished it from almost every other hotel I had ever been in was that its profits, beyond those required to run it, replace items stolen by the guests and make provision for the staff, went to support a school for poor children, a bequest by the late owner, a philanthropic Turk. It was thus the only hotel in which by simply staying in it, one could be accused of actively helping to do good.

32

A Journey in the Wilderness
(1971)

This journey, made in February 1971, to the Monastery of St Katharine in the mountains of the Sinai Peninsula, was one that had been difficult to arrange in the way in which we wanted, which was alone, apart from the escort the Israelis obliged us to take along. At that time a state of war existed between Egypt and Israel, and fighting was quite likely to break out again at any time in the course of which the monastery, together with its priceless contents, might well be destroyed – at the very least the monks might be forced to evacuate it, in which case it would probably be pillaged. At this period the whole of Sinai was occupied by the Israelis, the Egyptians having been driven back across the Suez Canal where they were now engaged in licking their wounds, pondering revenge and infiltrating guerrillas into Sinai across the Gulf of Suez.

On the journey through Sinai to the monastery we were accompanied by three Israelis, all armed to the teeth with Russian Kalashnikov rifles, captured from the Egyptians during the Six-Day War in 1967, and .38 Smith & Wesson Police Specials, the ones with the short barrels that had been made fashionable by Ian Fleming, whose Bond thriller, *From Russia With Love*, had a dust jacket adorned with a highly realistic depiction of one of these weapons.

The journey from Suez to the monastery used to take eight days by camel, with eighteen days being the minimum time recommended for the entire trip; but it could be done in five days. Two days by ship to El Tor, a quarantine station for pilgrims returning to Egypt from Mecca, which was situated on the west coast of Sinai about one hundred and fifty miles southeast of Suez, then three more days by camel to the monastery. Europeans were strongly advised to avoid making the return journey by ship and to travel instead by camel, as this would enable them to avoid spending two ghastly days in quarantine at Suez.

All this was in the now distant past, long before the Second World War. By 1971 it was no longer possible to travel to the monastery by ship, camel, motor car or any other means from Egypt, now that the Israelis had occupied Sinai. The only practicable route was from Eilat, the port the Israelis had built at the head of the Gulf of Aqaba, southwards along the almost waterless east coast of Sinai to a place called Sharm el Sheikh, near the tip of the peninsula; then up its equally desolate west coast to El Tor through a coastal plain which sloped up gradually to the foot of the high mountains of the interior. The track that ran through it had a specially prepared expanse of sand throughout its entire length on its seaward side, which was inspected by patrols and raked daily, so that any footprints left by the then numerous infiltrators who used to cross the Gulf of Suez from Egypt under cover of darkness could be detected.

El Tor by that time was a lonely place, deserted because even the poorest pilgrims now travelled by plane to Jiddah, the port of Mecca on the Red Sea, rather than by ship. There was a small inlet between reefs from the Gulf of Suez (frequented by the dugong or sea cow, whose skins were used for making native sandals), some jetties, some palm groves, some old rotting boats high and dry on the shore, of the sort once used in the pilgrim trade, some decayed buildings of unburnt brick, some mosques, some modern apartment buildings and the great wired-in quarantine station – eerie as such places invariably are, now all deserted, evacuated during the Six-Day War, the only signs of animation being the columns of whirling sand chasing through this ghost town on the wind.

From El Tor the track continued northwards, past Hammam Seiyidna Musa, the Baths of Our Lord Moses, where some hot springs issued from the base of a hill, past Jebel Nâkûs, the Bell Mountain, an expanse of yellowish-brown sand about two hundred feet high which was supposed to produce deep, swelling, vibratory, moaning noises when disturbed, but failed to do so on this occasion. Then, having reached the mouth of Wadi Feirân, we turned away through it to the east.

Wadi Feirân concealed within it the most beautiful and fertile oasis in all Sinai, where the most delicious dates and the best vegetables were grown. It was the abode of Sheikh el Sheikh Abu Abdullah, chief of the Umzeini Beduin and virtually king of the entire peninsula. It was also the site of a great cathedral monastery, now ruined, that had been the original episcopal see of the bishops of Sinai, the first of whom may have been named Moses, consecrated Bishop of the Saracens in AD 373 or 374. Above it rose Jebel Serbâl, originally identified, rather than Jebel Musa, as Mount Sinai, the mountain on which Moses received the Tablets of the Law. It was for this reason that Wadi Feirân became the seat of the episcopal see, and that the granite fastnesses of Jebel Serbâl and other neighbouring mountains are covered with the remains of chapels and churches and honeycombed with tombs and hermits' cells.

The approach to Wadi Feirân was uncertain country to wander in at that time. It was full of minefields, relics of the Six-Day War (which by now we were heartily sick of hearing about), some of which had been moved miles from where they had originally been planted by flash floods which happen very frequently in the winter months and which make a bivouac at this season a hazardous proceeding.*

And it was undoubtedly penetrated by infiltrators, for it was here that our Israeli bodyguard found prints of the sort of boots worn by Egyptian infiltrators. This discovery thoroughly put the wind up them, and set them cocking their carbines and inspecting their Smith & Wessons, although asking them to enlarge on the subject of infiltrators was rather like asking a nanny how babies are made.

It was soon after making this discovery, while travelling through a defile, that we met the harem of Sheikh Abdullah on the march – a convoy of camels, with the black, tent-like litters in which the women were hidden, on their backs – swaying down to us, escorted by a number of Beduin armed with

*On 3 December 1860, after little more than an hour's rain in Wadi Feirân, the torrent was eight to ten feet deep; an Arab encampment was swept away, and nearly thirty people, with scores of camels, donkeys, sheep and goats perished; but only two bodies were found, the rest being buried under debris or swept right down to the sea. *Ordnance Survey of the Peninsula of Sinai*, Part 1, 1869, pp. 226–7.

captured Russian carbines. There was also a very smart jeep, in which the Sheikh's veiled favourite sat in front next to the driver, with two more armed men in the back. Noticing the presence of a European woman in our party she gave orders that we were to halt, and taking Wanda some way off to the shelter of a cliff, where the two of them were beyond the prying eyes of men, proceeded to unveil herself, displaying dazzling charms.

Late in the afternoon of the second day of the journey from Eilat we were in Wadi Turfa, still well to the north of the monastery which stands at the foot of Jebel Musa, otherwise Mount Sinai, at the bottom dead centre of the peninsula, climbing towards it among a few turfa trees, a sort of tamarisk. In spring and early summer the twigs of these trees are sometimes punctured by an insect, and they then yield a gum tasting of honey, which falls to the ground where it coagulates. The Beduin used to collect these manna-like drops and bring them to the monastery, where they were packed in tin boxes and sold to pilgrims as the Manna of the Israelites. Perhaps they still did sell it to pilgrims. Nothing had changed much in this part of the world since the sixth century when the monastery was built, or since the Israelites had wandered here, for that matter.

At about four o'clock we came to a gorge, the Naqb el Hawi, the Cleft of the Wind. To the left there was a white-capped rock on which the Beduin believe Abraham performed his sacrifice, and another on which Moses sat while tending the sheep of his father-in-law, Jethro.

From this gloomy defile we emerged on to a long, wide, undulating plain flanked by high, jagged and ruinous-looking peaks that looked like heaps of baked and broken bones, in which the black tents of the Tuarah Beduin – Tuarah means 'mountaineers' – were pitched close in under them on the valley floor.

The Tuarah women (no men were visible) wore long black cotton shirts. They reached to the ground and were gathered at the waist and embellished with strings of white beads and embroidery. They had black kerchiefs on their heads edged with gold braid and over these they wore white, hip-length woollen mantles; from beneath the kerchiefs and the mantles their jet-black hair protruded, which was done up in what looked like buns over their foreheads. The red veils they wore were hung with coins, pearls and gilt or gold chains. When they saw us they drove their goats away as if we might contaminate them.

We came to a couple of whitewashed tombs on a low hill: the larger contained the remains of a notable saint, the Sheikh Nebih Salih, to which the Tuarah turn when they pray instead of towards Mecca, being pretty vague about Islam generally and living in a region where Christian and Islamic myths are particularly mixed up, one with the other. On his coffin stood a green silk turban, and some offerings: a tin-opener, some pieces of mirror, a plastic bag. At one time the tomb was embellished with more picturesque objects; ostrich eggs, shawls, halters and bridles; but no more.

At about five o'clock we came to the foot of a great wall of mountain, scoured with deep ravines and crowned with peaks, with other valleys stretching away on either side of it from the one in which we had now halted, giving distant prospects of other massifs, other peaks.

This great mass confronting us was Horeb, the northern part of what the Arabs call Jebel Musa, Mount Moses, otherwise Mount Sinai, the Mountain of the Law; and the plain below it, on which we had now halted, was the place where some savants and many faithful people believe that the assembled Children of Israel, about two million of them, stood when the mountain was wrapped in smoke, the Lord descended on it in fire, and the whole thing shook to the accompaniment of a prolonged trumpet blast. Seeing it at that moment, with the mountain rearing its shattered peaks above the darkening, desolate plain, it was easier to believe than not to believe that these events had taken place where we were now standing.

The sun was gone from the valley now, except from the summit of a little hill with a Christian chapel and a Muslim shrine standing together on it. This was Jebel Harun, Aaron's Mountain, on which he set up the Golden Calf while his leader was on the mountain above. Here, round about Mount Sinai, practically every site hallowed in the Old Testament is also venerated by Muslims (it would be difficult to overstress the importance which the Prophet accords to Moses and Mount Sinai and which is expressed in the Koran).

Then having arranged with a young Jebeliyeh, one of the serfs of the monastery, a boy of markedly European appearance, dressed as a labourer rather than a Beduin, to bring camels at four a.m. for the ascent of the holy mountain, we rounded the hill into Wadi Shuaib; and there was the monastery, with the great granite mountains, hot-looking even in shadow, looming over it, those on the left with crosses set up on them in inaccessible places, those on the right outriders of Mount Sinai, exactly as it had been in all the nineteenth-century engravings I had ever seen: a miniature city with a bell tower and a square-towered minaret showing above the immense stone walls which enclosed it, and outside in an enclosure, a small dark forest of cypresses.

By this time it was almost dark. In another few minutes the outer gate and inner doors would be closed for the night, and we would be benighted in Wadi Shuaib. Until the middle of the last century there was no outer gate and there were no doors. Therefore any visitor, having first sent up his letters of introduction or other credentials for inspection in a basket was, if found to be *persona grata*, wound up more than thirty feet in the air into the monastery by a wooden windlass, through a hatch grasping a greasy rope and sitting on a wooden bar attached to the end of it, in much the same way as a visitor to some of the monasteries of the Meteora in Thessalia on mainland Greece used to reach their objectives.*

Now, instead of being hauled aloft, the driver was sent in through a side gate to negotiate for rooms, which had to be paid for in advance, and when he returned he grumbled a lot because it cost ten Israeli pounds a head, which to me did not seem excessive in what was in fact the very heart of the Wilderness. Over the centuries the monks have always had difficulty in getting rid of pilgrims and other visitors, especially those already

*There was a main gate to the monastery, but it was walled up and only opened to admit a newly elected and consecrated archbishop. This practice was discontinued in 1772, and from 1782 until 1872, whether newly elected or not, no archbishop visited the monastery at all. The winch has (or had) the date 1791 and the monogram of St Katharine carved on it, although the method of entry dates from a much earlier period, being first mentioned in 1512 by Jehan Thénaud.

in residence whose credit was exhausted, justifiably aghast at the prospect of being turned loose penniless and provisionless in such barren regions.

We unloaded our gear and went in through the outer postern and three more studded iron doors in the main wall which, before the outer gate was built, led only to the gardens and the charnel house, and up past the mosque built at the beginning of the twelfth century, to the part reserved for visitors in the north-west angle of the monastery. No one is quite sure why this mosque was built, although it must have been an advantage for Christian monks to have a mosque on their premises in times of trouble with the Muslims.

Whatever the reason, when the Khedive Abbas, ruler of Egypt from 1849 to 1854, was ordered to Sinai by his physicians for the good of his health and lived in the monastery, while having a palace built on a nearby mountain top which he never occupied, he prayed in the church, never in the mosque.

We drew bed sheets (having pointed out that we did not relish the ones in our cells which bore the impress of what must have been numbers of previous occupants) from a distinctly worldly, if not depraved, monk of about thirty-five who looked as if he observed the feasts of the church rather than the fasts but was nevertheless an amusing fellow. We then visited the kitchens where we handed over a number of cans of food to a Jebeliyeh cook (visitors had to provide their own victuals). He immediately immersed them unopened in a huge cauldron of boiling water in which lumps of some unidentifiable meats were already seething, where a number of them eventually burst.

By the time everything was sorted out it was almost the hour for lights out. Electric power was produced only in the evenings by a generator which closed down at nine p.m. sharp. At ten to nine there was a premonitory flickering of bulbs, upon which everyone – monks, guests and any Israeli soldiery immured there on outpost duty (the monks were regarded as a security risk, presumably because their archbishop resided in Cairo) – began to zoom around, some of the most fortunate wearing long woollen underwear, all trying to reach their respective 'dorms' (dormitories) before total darkness prevailed. Although the monks manufactured candles from beeswax they were not available for secular purposes. It was early February, and we were five thousand feet up in the Sinaian highlands. What we needed were arctic sleeping bags, some of the long woollen issue underwear and portable stoves. Without these, nights in the monastery were almost unimaginably cold. The summers, we were told, were almost equally unimaginably hot, but summer was months away.

The last thing I saw before the lights cut out was a battered notice: 'Suez to the monastery two hundred and eighty miles. Cairo to the monastery four hundred and seven miles'.

The monastery is Greek Orthodox. It was founded in AD 530 by order of the Byzantine Emperor, Justinian I, on the site where the Burning Bush still grew, in which God appeared to Moses. There was already a well of good water and a tower in which the hermits and the other holy men in which the region abounded, took refuge from time to time in order to avoid being slaughtered by the Saracens, although they were seldom successful in this.

When the commissioner responsible for the construction of the monastery returned to Byzantium the Emperor asked him why he had built it in a place where it could so easily be rendered untenable by the barbarians, who had only to stand on the slopes of the mountain on the eastern side and lob rocks into it (he seems to have forgotten about the necessity of building it around the Burning Bush). The commissioner's excuse, that there was no room for such an edifice on the adjacent peaks and that there would have been trouble with the water supply, was not regarded as sufficient by the emperor who, before having him beheaded, told him that he should have reduced that particular mountain to the level of the surrounding plain before proceeding with the work.

The Emperor provided the monastery with a staff of slaves, some of whom were Egyptian, some Christians from Wallachia (a region of southern Roumania), and their descendants still performed the menial tasks of the monastery. Some of them are supposed to resemble Wallachians to this day, but if one has never seen a Wallachian it is difficult to hold any strong opinion about this – the Jebeliyeh boy I had met the previous evening

266

was certainly more European-looking than Egyptian or Beduin. The last Christian Jebeliyeh, a woman, died in 1740, but although they are now all Muslims they are still looked down on by the Beduin of the peninsula as Nazarenes (Christians) and *fellahin* (peasants).

These Jebeliyeh, People of the Mountain, as they are known, worked in the gardens, at the monastery and also in Jebel Feirân, where they cultivated dates. They also provided camels to carry visitors part of the way up Mount Sinai. For this they received a small pittance from the monks, who reserved to themselves the right to strike all bargains with visitors. The monks also provided them with their daily bread, which was baked in the monastery and then let down to them by the windlass. Although a custom hallowed by age and of considerable importance to the Jebeliyeh, it was not a very edifying spectacle, and it was said to have been interrupted by the Six-Day War.

It was for this reason that the camels, which had been ordered from the Jebeliyeh for four o'clock the following morning, failed to appear. At the time, being ignorant of the hierarchical arrangements, I attributed this to the fact that it was still pitch dark and freezing, not because I had failed to book them through the proper channels. As a result several hours passed before the gates were finally opened, and when we did begin the ascent of Mount Sinai, it was on foot by way of the camel track, one of the five accepted routes to the summit.

On the summit there was a chapel with a corrugated-iron roof and a mosque, both built of the pink granite of which the mountain is made. Under one side of the mosque there was a sort of cave, nothing more than a hole really, in which Moses crouched while the Glory of the Lord passed by during his forty-day fast on the mountain, now filled with the debris of picnics. Here, once a year, the Muslims sacrificed a sheep or a goat to the memory of Nebih Salih whose tomb, or one of his reputed tombs, we had visited on the way to the monastery, smearing the doorposts of the mosque with its blood.

There was not much room for anything else on the summit of Mount Sinai. From it, in the screaming wind which had turned the granite slabs which littered it to plates of ice, there were awesome views to the Red Sea, nearly seven and a half thousand feet below, over some of the wildest and most desolate country on earth. These views included Jebel Katharina, which is more than eight and a half thousand feet high. It was to this mountain that the remains of St Katharine were transferred by angels from Alexandria in the early part of the fourth century, after the machine specially constructed to torture her had chewed up its operators instead, and there was no one sufficiently skilful left to do anything but behead her. There on the mountain, in the eighth or ninth century, the monks found the body of the young virgin, unmutilated, uncorrupted and resting in a rocky depression. The depression was full of what was to become much venerated and much prized oil, which her body had secreted, and it continued to do this in gradually decreasing quantities until the flow ceased altogether in 1489, by which time it was reduced to three drops a week. From the summit of the mountain they carried her down to their monastery, which up to this time had been dedicated to the Virgin Mary and henceforth would be dedicated to St Katharine.

We went down by the old penitential route, which was designed to be ascended rather than descended if one was to appreciate properly its penitential significance. It consisted of thousands of stone steps, variously estimated as being fourteen thousand by an Italian pilgrim called Gucci in 1384; six thousand six hundred and sixty-six by an Arab, Al-Makrizi, who lived from 1364 to 1442, quoting another Arab; three thousand five hundred, source unknown; and five thousand by Richard Pococke, the English Orientalist and traveller who visited the monastery in 1726. However many there were, and we soon gave up counting them, it must have been an appalling job putting them in position.

On the way we saw the footprint of Mahomet's camel and three chapels, one dedicated to Moses or Elishah, another to Elijah, and the third, the Chapel of Our Lady Oikonomissa (on the Mount), to the Virgin Mary. It was here that together with the Child Jesus the Virgin made a miraculous apparition, which led to the monastery being supplied with a camel train of food, and to the banishment of a horde of giant fleas which had infested it so thoroughly that the monks had been forced to evacuate it. We also passed through two penitential gateways, at one of which St Stephanos the Doorkeeper used to sit hearing confessions or receiving certificates from pilgrims attesting that they had already confessed and were therefore in a fit state to ascend the holy mountain. At the very bottom was the Well of Jethro, where Moses is said to have watered his father-in-law's flocks, although the monks themselves believed that its only claim to fame was that Sangarius, a cobbler who was also a saint, drank the water from it when he lived there.

At the time when I visited the monastery there were only eight monks, most of them Greeks and Cypriots, all members of the order of St Basil. In the year 1000 there were about three hundred; in the fourteenth century the numbers varied between four hundred and two hundred and forty; in the fifteenth and sixteenth centuries there were between thirty and fifty. In 1620 there were only two or three. In the eighteenth century the numbers rose again to around fifty, but in the nineteenth century they were down to less than thirty. By 1938 the actual inmates were reduced to nineteen. (The total number of monks in the whole of Sinai, at that time, including those working outside the monastery, was forty-nine.) In 1971 the oldest monk was eighty-five years old and had been in residence for forty years; another, also very old, for twenty years. Three of the eight monks who made up the entire population of the monastery, the Vicar-General, the Treasurer and the Steward, constituted the chapter, which discharged the obligations of their 'Father-in-God, the Archbishop of Sinai, His Beatitude Gregoric II'. They did so according to a precise set of rules as constricting for the archbishop as they were for the monks and chapter. At that time, with a state of war existing between Egypt and Israel, the archbishop spent three months each summer at the monastery, and the remainder of the year in Cairo, travelling between the two by way of Cyprus with the aid of a special *laisser-passer*.

The longer I remained within the walls of the monastery, and every day I became more disinclined to leave it, the more striking its resemblance to Mervyn Peake's mythical Gormenghast, which he describes in a series of novels, became. Within its walls, just as in Peake's enormous castle, a handful of persons, not all of whom could be said fully to understand its significance (I intend this in no derogatory sense), sustained a highly complicated, protracted and exacting ritual, beginning the first of their twice-daily devotions soon after four in the morning – at one time when there were more monks they took place twice daily and twice nightly – summoned by the striking of the *symendra*, a resonant piece of wood, having been previously awoken by the tolling of a bell to the number of times equal to the number of years of Christ's life on earth; but unlike some of the inhabitants of Peake's demesne, living on what was a very thin diet for anyone engaged in all the multifarious activities which the monastery demanded in order to keep it going.

Beyond the walls lived the equivalent of Peake's wood-carving 'Dwellers', the Jebeliyeh, serfs dressed not as Beduin but as *fellahin*, with whom the inhabitants of the monastery had no kind of social intercourse, and of whom very few were ever admitted within its walls, and then only to perform menial tasks. As in Gormenghast there was a library, to which probably not more than one or two of the occupants had recourse or even access. It was one of the oldest and richest libraries of its size in the world, and one of the least studied, made up of more than three thousand Greek, Arabic, Syriac, Palestinian, Georgian, Armenian, Slavonic and Ethiopic manuscripts, including the Codex Syriacus, parts of which date back to the fourth century and earlier.

Peculiarly Gormenghastian are the circumstances in which Constantin von Tischendorf saved forty-three priceless leaves of the Septuagint from being incinerated out of a total of one hundred and twenty-nine which still existed in the monastery in 1844, and of how, in 1859, he was able to return there, after an abortive visit in 1853, and acquire three hundred and forty-seven leaves of a fourth-century manuscript of the New Testament and parts of the Old Testament, which was to become known as the Codex Sinaiticus, thus preserving them from almost certain destruction. Yet in 1971, more than a hundred years later, Tischendorf's name was still anathema in the monastery.

Extraordinary, too, were the congeries of buildings linked by a labyrinth of passages. Exploring them was like a dream in which one floated through narrow tunnels in the immensely thick walls (the lower courses made up of enormous blocks of granite), constructed by Justinian's men and repaired by those of General Kléber, sent there for this purpose at the time of Napoleon's expedition to Egypt, with on either side the workshops of Jebeliyeh craftsmen dead and gone, who also worked on the fabric; past chapels closed because there was no one to serve them any more; up winding staircases; along balconies supported by flimsy bamboo laths and crumbling plaster; into vaults and windowless courtyards, in which the apparatus for distilling raki from dates lay long abandoned. (Apparently the principal reason for this distillation was to gain income by selling it to pilgrims. Records of monkish drunkenness are rather rare.)

On under trellises of vines, into bakeries furnished with wooden moulds to embellish the bread with the outlines of St Katharine and huge wooden troughs excavated from trunks of trees (How did they come here, and where did they come from? Were they dragged overland from Lebanon?); into a disused refectory with a vaulted roof and embrasures cut through it to let in the light, its walls covered with graffiti done when this was, temporarily, a Crusader officers' mess. Outside, growing against one of the curtain walls in a little enclosure, was a plant that resembled a raspberry. A notice, gratifyingly in English, stated that it was the Burning Bush. It seemed impossible that all this could exist within a space of 280 by 250 feet.

A door was opened and we were confronted with the skulls and bones of three thousand monks, neatly stacked. How little space three thousand of us take up when our skulls and bones are neatly stacked. Others, abbots mostly, were alone in small crates. And presiding over this charnel house, dressed with ghastly elegance, almost gaiety, in what the monks called megaloschemos, the 'robe of angels', worn only by monks of the highest monastic rank, were the remains of St Stephanos the Doorkeeper, who used to sit hearing confessions at the Penitential Gateway on the steps leading up to the holy mountain.

The most important remains were in the Church of the Transfiguration, the Byzantine basilica: the skull of

St Katharine, covered with jewels, and her left hand, each contained in a golden casket. In the church was a small part of the more than two thousand ikons, the earliest of which date from the sixth century, with which the monastery was endowed. And somewhere in the sacristy of this church was the treasury which only the treasurer, not even the archbishop, had the right to open. More than two thousand ikons, some of them painted in the sixth century, together form what is probably the most important collection in the world.

In the library was the Hand of Mahomet, outlined on a document in which he gave security to the monastery and its occupants for ever. It was said to be the copy of a forgery. Whatever it was, the monastery and its monks were still miraculously intact in what was one of the lonelier Christian outposts, and at the time of writing they still are.

Outside in the 'real' world, not more than an hour's brisk walk for an Englishman along the foot of Mount Sinai, was the place where the earth opened and swallowed Kora, Dathan, Abiram and their followers; the rock which Moses smote and from which the water gushed; the mould which Aaron used to cast the Golden Calf; and the place where Moses broke the Tablets of the Law. Whether it all happened here, on and around Jebel Serbal, or at Petra in Jordan or in Saudi Arabia, is not of very much consequence. There, in the heart of Sinai, with such a wealth of terrestrial evidence, the Old Testament came very much alive.

33

Wimbledon to Italy by Bicycle
(1971)

Twice a year we used to go to Italy to see Wanda's mother in the Carso and to work a vineyard in Northern Tuscany. In early spring we dug and manured it and in the autumn we made the wine. In September 1971, having once more to go to Italy for the *vendemmia*, the vintage, and fed up with racing across France by car without ever having time to see anything worthwhile en route, I decided to acquire a bicycle to ride to Italy from Wimbledon, where we lived, following canal banks and other pleasant, vegetable routes through France. This gave me the excuse to order a bicycle. The sort of bicycle I ordered was the equivalent of a *gran turismo* motor car; fast and comfortable over long distances. It was also supposed to be a machine that could have any of its component parts replaced in France or Italy without having to wait about for days for items to be sent out from England. I only had ten days to spare for this journey and I reckoned that I would have to cover at least 1250 kilometres between Wimbledon and Alessandria in Italy, which for me was the end of the road so far as cycling was concerned. Cyclists are not allowed on *autostrade* or any other sort of motorway, and I had no desire to ride along the Via Aurelia, the coast road from Genoa to La Spezia, which is highly dangerous, full of *autotreni*, huge lorries with trailers, and which has on it, as an ultimate deterrent, the atrocious Passo di Bracco. Neither did I have the time to wander through the Apennines on lesser, quieter but immensely mountainous roads, none of which would have delivered me where I wanted to go.

When I went to collect the bicycle in the Midlands where it was made, it seemed like a Euro-marketeer's dream. The chain-wheels – you had to have double ones for anything in excess of five speeds and I had been persuaded that I needed ten – the cranks, pedals, bottom bracket and head assemblies, fork-ends and seat pin and the gear changing mechanisms were all Italian, made by a firm called Campagnolo, and unbelievably expensive. The hubs and multiple free-wheel were French; the alloy wheel rims and the brakes were Swiss and the tyres were Belgian – almost unbelievably, by 1971 Dunlop had given up making bicycle tyres in Britain. What was left, or most of it – the frame, made of Reynolds 531 butted tubing which even the most xenophobic of continental riders regarded as good, the mudguards and pump, the leather saddle, the handlebars and the handlebar extension, were all British. I never found out where the chain was made. There was no earthly reason why that should not have been British too. Reynolds, a British firm, still made the best chains in the world and not only for bicycles; but by the 1970s everyone had gone mad about French and Italian accessories (British-made bicycle frames, allegedly French, were actually being sold to the British under Italian and French names), and as a result many British accessory manufacturers were giving up.

I took camping kit with me, which was a mistake as it took ages packing up each morning, and eventually towards the end I slept indoors: a 3¾-lb tent, a 3½-lb sleeping bag, a canvas water bucket and basin, and a Meta stove for making tea – I always ate in cafés and restaurants as cycling to Italy in ten days provided me with quite enough exercise without cooking and washing up.

What brought, together with changes of clothing, the loaded weight of the machine up to a staggering sixty pounds were the tools for a bicycle that, it turned out, was partly constructed in English feet and inches and was partly metric. To do anything to the Campagnolo bottom bracket a number of very expensive tools were needed, and special spanners and a spring-loaded instrument were desirable for adjusting the brakes. Even with all this I somehow failed to acquire what turned out to be a small but vital piece of equipment, known as a free-wheel block remover. I also had to carry spare spokes, brake blocks, inner tubes, brake and gear cables, batteries for my bicycle lamp. I was told that one could not buy battery-operated bicycle lamps in France or Italy because they only used dynamo lamps, which operated off the tyre and which were not much use in a tent, unless you had a slave to turn the wheel. I also took a candle, nine Michelin maps and two Touring Club Italiano maps for Italy, as I could not be sure that I would be able to buy the next sheet en route when I ran off the previous one. I also started off with some excellent green Michelin regional guide books but I could not face carrying them and gave them away.

Day 1

A mysterious distortion in the wired-on tyre on the back wheel developed two hours out of Wimbledon, giving the illusion that the wheel was buckled. Believing that there would be enough time at Newhaven to buy a new one, I pushed on. At Newhaven (eighty kilometres) I had to choose between buying a tyre or missing the boat, but I comforted myself with the thought that as the bicycle had 27-inch wheels, a size I seemed to recall that had been used in France before the war by racing cyclists, I would have no difficulty in replacing it.

However it was not so. 'You will not find a shop with *une enveloppe anglaise* in all France,' said the proprietor of the best bicycle shop in Dieppe, with what I identified as Gallic relish. 'Your *enveloppe* has an ineradicable defect.' And how long it would last before it collapsed was anybody's guess – thirty kilometres, a hundred . . .

Apparently French and Italian bicycles fitted with wired-on tyres, as opposed to tubular tyres which are stuck on, now had slightly larger diameter rims than British ones. The best thing, he said, would have been for me to have had my bicycle fitted with tubular tyres which were the same size in Britain as on the Continent, but they needed a different sort of rim and, anyway, such tyres are more suitable for day trips or touring with minimal luggage than for cycle-camping with a comparatively heavy weight over the back wheel.

In the face of all this depressing news I was nevertheless reluctant to return to Newhaven and face two more sessions with the French ship's gruesome ham sandwiches and equally gruesome self-service cafeteria – by this time, the early 1970s, the French were catching on fast to what the British had known for a long time, that it was not necessary actually to provide any sort of civilized service at all on a cross-Channel ferry service, as even if you offered them nothing the customers would travel just the same. I therefore decided, stubbornly and irrationally, to press on to Rouen and try the Michelin depot there. Had I had any sense I would have telephoned them, but even if I had done so they would have told me that they had the size I needed in stock, which was not in fact the case, so the result was the same.

4 p.m. The Route Nationale unrolled ahead of me between the enfilades of poplars, like an endless strip of paper. How vast France was. By the time the environs of Dieppe were left behind, cars with GB plates were already thin on the ground. Soon they disappeared completely and I was alone with the Citroëns, the long-distance lorries that seem to coast past at a hundred and fifty kilometres an hour, and strange, bus-like but windowless, pale-grey vehicles, containing what – the guillotine? Not only empty of vehicles, empty of French, except in the towns which in France are so evenly spaced that they could have been established where they stand by ministerial decree, towns that have things we do not have in Britain, or if we do are not the same: *charcuteries, drogueries, huissiers, terrains viabilisées, toutes directions*, signs which I now knew, after years of being taken in by them, despatch you where you do not want to go (the only way to deal with a French town is to charge through the middle of it), past devilishly-sited *priorités à droits*, from which old ladies in rusty black shoot out on autocycles, like witches on broomsticks. '*Faites attention!*' If the French say something they mean it. Devil's Island was established expressly for those who do not believe in *la loi*. Which was why I soon left this Route Nationale to travel on 'D' roads, *chemins départementals*, which are generally much safer for cyclists than English roads and often go on and on across enormous tracts of country avoiding all but the smallest towns. 'V' roads, *chemins vicinal*, are quieter still.

Spent the first night in the back garden of the Auberge du Val du Saâne, in the pastoral valley of the Saâne, having dined in it and having covered one hundred and ten kilometres on my bicycle since leaving Wimbledon.

Day 2

At Rouen great excitement when a Michelin man announced, after a good rummage, 'Yes, we have twenty-seven-inch *enveloppes.*' Removed the rear wheel – not easy as you cannot turn a bike upside down with loaded pannier bags fitted to it, and these particular models took ages to put on and take off – then removed the tyre only to find that whatever his *enveloppe* was it was not twenty-seven-inch.

The Michelin man went off to telephone Paris and eventually returned with two interesting alternatives, both almost equally awful. Either to spend the weekend in Rouen – this was a Friday – and await delivery of a $27'' \times 1\frac{1}{4}''$ *enveloppe* some time on Monday, or somehow get to Clermont-Ferrand in the thousands of feet high Massif Central, a region I had been planning to avoid at all costs, only about five hundred and fifty kilometres to the south by Route Nationale, but presumably much more by the kind of roads I used. There, he said, there was a Michelin *usine* which, as a rare example of French clemency, turned out these *mini-enveloppes* for the British who were mad enough to use them when almost no one else in Europe did, except perhaps the inhabitants of Gibraltar.

After thinking about this while eating a cheap, copious but rather greasy luncheon for such a hot day at Au Carreau des Halles in the port area of Rouen, which was destroyed during the war and has been rebuilt in a manner that no one could describe as picturesque, I set off for Clermont-Ferrand, hoping that the tyre would hold out, but with all my romantic visions of cycling day after day along canal banks under the plane trees, which was what I had planned, now dispelled. In doing so, in order to avoid going through Rouen itself, which is in-conveniently situated so far as cyclists are concerned in a hole in the ground, I made what turned out to be an unwise detour down the right bank of the valley of the Seine, large parts of which are an industrial mess, pedalling past Flaubert's Pavillion at Croisset, now a museum, which stood below steep chalk cliffs, both of which, museum and cliffs, looked as if they could have done with a good rinse. I then crossed the river at Val-de-la-Haye in a motor-boat to the left bank where I immediately got lost, first in a forest which had been messed up by the builders of the Paris-Caen *autoroute*, then in a labyrinth of signpostless lanes from which I was rescued, almost weeping with vexation, by a kindly lorry driver.

7 p.m. Reached St André-de-l'Eure, having ridden one hundred and twenty-eight boiling kilometres and having passed through Evreux in the six o'clock rush hour – never again on a bicycle! Ate a *prix-fixe* dinner, *rilettes* (ugh, in such heat!) and *tripes*, separately of course, at the Café de la Ville, to which, in spite of it costing only F8.50 [64p or $1.60], I shall not be returning. [At this time, in 1971, sterling was about F13.30 to the pound.]

Camped in the midst of huge, flat, prairie-like fields and, as it was very early and still daylight, thought about France and the French, about which I now thought I knew a little more than I had in 1946, up to which time I had never been to France, and made some notes about them which were eventually expanded into what follows. (Any reader not interested in the writer's views on France but still keen on cycling can stop here and start again at Day 3.)

The emptiness of France is not a figment of the imagination. This is a country nearly four times the size of Britain, yet with a smaller population. In the villages, apart from one old man in faded *bleus* (dungarees) gazing at what to him (and to me) is agricultural machinery, the only other figure in sight is often the *poilu* on the war memorial. The huge, prairie-like fields, such as the one in which I am sitting in my tent, are as empty as any real prairie, except perhaps for one man with a tractor who often works far into the night using headlights (as is now common in other parts of Europe, including Britain). Yet this is, as it was before the tractors came, which is only recently, the most productive agricultural country in Europe.

The sun is setting now. It is the moment recorded in Millet's *Angelus*, but without the peasants with their heads bowed, a picture of which my Auntie May had a reproduction in her home in Stamford Brook, and now usually without the bell.

Out there beyond the *plaine* is the rest of France, a country arguably – and I am thinking of it from the point of view of a visitor, rather than an inhabitant of the Lens coalfields or of a workers' housing complex in Marseille – the most beautiful, in its infinite variety, in Europe.

Out there, too, are the French, a nation made up of Celts, Latins and people of Germanic origin: yet all of them regarding themselves, not as the Scots, Irish and Welsh tend to, and increasingly so, as separate, distinct nationalities within the British Isles, but as one people, wholly and utterly French. A people who in moments of collective emotion may begin to sing *La Marseillaise*, as the French prisoners-of-war did on hearing of a French victory at Verdun, in the film *La Grande Illusion*. To the British, the thought of singing 'God Save the Queen', admittedly an anthem with less verve, except on strictly ritual occasions, or before the first act, would be unthinkable. If one's ship was sinking one would think twice about singing even 'Rule, Britannia'.

A country with a working class which has largely fled the land, leaving a train of deserted or semi-deserted villages, as any visitors to France can see for themselves. Many of them succumb to what is known as *la Tristesse Ouvrier*, a malady described by one French writer (Georges Navel) as 'a kind of pervasive depression induced by the claustrophobia, monotony, fatigue and insecurity of factory labour and by a continuing nostalgia for country ways'.

A country with a capital which in spite of years of what has been called *gaullo-destructiomanie* still contrives to be the city of which Flaubert wrote, 'an ocean in which there will always be unexplored depths'. A city in which a *toubib* is a doctor, a *pote* a friend, a *panier à salade* a police car.

The French, whatever changes are taking place in their way of life, are still a people who believe in work well done, in craftsmanship, which has its origins among the peasantry. They have a genius for the production of prototypes, what the *hauts couturiers* call *modèles* or *toiles*; less interested until recently in the dissemination of copies *en masse*.

A people of infinite resource. They have given birth or are popularly supposed to have done so, especially among themselves, to the aeroplane, suppositories, the submarine, Colette, the soufflé, Chanel Number Five, Wagons-lits (a wagon-lits is a wagon-lits, even if its owners insist on calling it a *schlafwagen* or a sleeping-car), Château Yquem, Tintin and Milou and the Vuitton trunk.

The French are still capable of writing in Michelin in terms of self-congratulation of their best restaurants as no other people in the world would dare, or have the right to, except the Chinese, who resemble them in their contempt for foreigners and in the intimate solidarity of their family life.

A people with the oldest colonial tradition in Europe who built up one empire, lost most of it and then started all over again to build another, the second biggest in the world, in which no one in what the empire builders called La France de la Métropole was ever very interested. Voltaire's description of Canada as a 'few square miles of snow' was not only witty but typical of what les Métropolitains felt about it. In the thirty years since the war I had followed many of the trails that they had left: in the New World where they had opened up the country from the Atlantic to the Rockies and as far south as New Orleans and the Caribbean, on the feverish coasts of South America, in Syria and Lebanon, where descendants of Scandinavian pirates who had remained long enough in La France de la Métropole to become assimilated before taking off for Hastings, Sicily and the Levant, had left their mark. And I had seen the remnants of their stations on the Hooghly, at Chandernagore, and on the coast of Coromandel, at Pondi-cherry, where Indian policemen in *kepis* still trilled away, just as if it was half past five in the Place de la Concorde.

I admire them, but like so many of my fellow islanders, I find it difficult to be friendly with them. They are too prickly.

By this time it was dark.

Day 3

Set off at 6.30 a.m. Huge moon above, freezing eye-level mist below. Descended into the beautiful valley of the Eure and followed it to its junction with the Aunay, a lesser stream. On the way discovered that I had left a satchel containing all my money and travel documents on a fence in a village while adjusting a brake. Went back five kilometres and found it still hanging there. Then followed the Aunay almost to its source emerging on another prairie, the Plaine de la Beauce on which the name of almost every village terminates in '-ville'. Memorable luncheon in a fantastic, untouched peasant interior, La Croix Blanche at Oysonville, about nine francs. Afterwards slept in a ditch – it was terribly hot – and then plunged into the Fôret d'Orleans, eighty-five thousand acres of beech, elm and ash, something which, thanks to the Forestry Commission, we will never see the like of again in our islands, and travelled through them down interminable grassy rides that were like something in a beautiful dream. Stayed the night at the Hôtel du Châteauneuf-sur-Loire which was cheaper than it sounded, having covered one hundred and sixty-eight kilometres. How much nicer French hotel and pub keepers – even quite grand ones – are to cyclists than many of their British equivalents met last year cycling across England.

Day 4

Crossed the Loire in freezing weather and rode upstream on the embankment towards Sully where I bought a large wound-dressing in a *droguerie* for my bottom, which now had more than one hole in it, with blood all over the place; and which, without a mirror, I stuck on with some difficulty behind a hedge, before riding south along the edge of the Collines Sancerrois. The sun was hot now: had a delicious mini-picnic under a tree, small, ripe, melons and a bottle of cold Sancerre.

7 p.m. Arrived at Dun-sur-Auron, south-east of Bourges in the department of Cher, having covered one hundred and thirty kilometres, just in time to have a hot shower in the municipal bath-house down by the river near the camp site where I then pitched my tent. Excellent dinner at the Hôtel de la Poste, Patron, M. Bresse, F8.50 plus wine.

Day 5

Left at 6.30 a.m. As always cold early, hot later. To Charenton along the edges of the Bois de Meillant, another noble forest. Ghostly white druidical-looking cattle – I suppose they're Charollais – in the fields, swathed in fog. Breakfast at a woodman's cottage in the great Fôret de Tronçais – home-cured ham, wine, freshly-baked bread and fresh butter, served by handsome, slim woman who wanted to come with me to Italy. What is the French for tandem – *tandem?*

Next, terribly hard hilly stretch through the outer defences of the Massif Central to La Roche Bransat where I lunched *chez* Madame Meunier, ideal place and food for tired cyclists – fresh fried eggs and tomatoes with herbs, steak and dollops of creamy potatoes, three kinds of cheese, fruit, a litre of wine, coffee, ten francs. She also had *chambres*. Pity it wasn't evening. Afterwards again slept in a field, ostensibly to dry tent which every morning I have to pack wet because of the heavy dew. Then wasted a lot of time trying to avoid cycling on terrible N9 to Clermont-Ferrand. Failed to do so and ate marvellous home-made ice cream at Gannat. The heat was appalling. Although I carried cold water in three thermic bottles they were so heavily insulated that there was hardly any room in them for any water. During these days I was doing about thirty kilometres to the bottle of beer or else to a *bière au pression* which was much better – cycling under such conditions is expensive. Finally left the horrible N9 at Aigueperse and arrived at dusk at the Hôtel de la Gare, St Beauzire, having covered one hundred and sixty-eight kilometres by my cyclometer. There was no longer a *gare* or a *chemin de fer*, but there was a stone trough in the back garden full of cheerful, naked *routiers* soaping and washing themselves all over, and I joined them.

No room at the inn, but ate copious, excellent lorry man's dinner for ten francs, and drank copiously, too, before pitching tent in the orchard; bombarded all night by falling apples. Hotel had a weighbridge and when they put my bike on it the scale turned at ninety kilos – can this be right?

Day 6

Cycle fifteen kilometres to Clermont-Ferrand to find Les Usines Michelin hidden away in vast Zone Industrielle. Courteously received. Bought three tyres and three tubes – not going to be caught again. Then they whizzed me in a van to the shop of M. Thomas, crack cyclist who had just returned from cycling to Czechoslovakia and whose hand-built bicycles were works of art. Here M. Thomas adjusted the *enveloppe* with his own hands. Saw from the map that because of having come to Clermont-Ferrand I could never now arrive at Alessandria in four days. It was over four hundred and twenty kilometres by way of the Massif to the Col de Larche where I planned to cross the Alps into Italy – this meant three days to Gap and probably another three to Alessandria, it was difficult to say. After agonies of indecision and disappointment I decided to take the train to Valence, and from Valence to Gap.

At the station at Clermont-Ferrand the staff tried to make me remove all the luggage from my bicycle before accepting it as accompanied luggage. I was lucky to travel on the same train with my bicycle – a rare event in France and Italy where not all passenger trains have luggage vans and the bicycle either goes on ahead or arrives later. For this reason, travelling with a bicycle in France by train, it is essential to have pannier bags with connecting carrier handles so that you can carry them, otherwise you are left on the platform with a mound of immovable impediments.

Dark when I got to Valence, a nightmare place in what was still apparently the holiday season, in the valley of the Rhône. Worse even than Montélimar – the town of nougat further downstream – where even the door knobs are sticky with the stuff. Town full of what the French call *cars* and I call enormous coaches. In the holiday season it is impossible either to eat well or sleep cheaply in such places, and probably at all other seasons as well. At the Restaurant des Gourmets I ate the worst meal I had ever eaten in France – significantly I was the only customer; and at the Hôtel des Voyageurs, on the advice of Michelin, I slept in a rotten room that cost a monstrous F37.40 [£2.80 or $6.70].

Day 7

To Gap up the beautiful valley of the River Dôme, passing the source of it en route. Wished I had had time to ride this section but although it was not difficult and there was very little traffic it was more than one hundred and fifty kilometres uphill and would have taken me at least a day.

At Gap I got my bicycle back and rode past an enormous, unlovely, artificial *lac* – the shape of holiday places to come, formed by damming two rivers, the Ubaye and the Durance, and covered with *amateurs des sports nautiques*. Then up the valley of the Durance to Mont Dauphin, an astonishing fortress, a little world really, as remote as the Potala, built of pink marble and set high above it on a vast, flat-topped rock. Very late lunch at a restaurant next door to the station at La Durance, in the valley below. F12 [90p or $2.00]. Delicious trout.

Now began the twenty-seven-kilometre climb to the Col de Vars in the lowest gear, thirty-six inches, which is not really low enough for this long, in many parts more than nine-degree ascent which is part of the Tour de France, but about the lowest obtainable at that time with a Campagnolo gear. Laden with camping gear I really needed a thirty-inch gear; but finally I wound my way up the endless hairpin bends to St Marcellin-de-Vars, a village among stands of larch with great snow-covered *pics* looming up beyond its pastures. Stayed at the Auberge Vieille which was as old as its name implied – and very cheap – where I was given a marvellous welcome. They washed all my clothes and then dried them overnight in a red-hot cellar.

Day 8

Left at 8 a.m. Gruesome stretch to St Marie-de-Vars through terrain ruined by winter-sport developments, which look even worse in summer when denuded of snow. Then through a wilder nine-degree stretch up to the Réfuge Napoléon which looked genuinely refuge-like and then to the Col itself (2109 metres) with the huge, sawn-off Brec-de-Chambeyron (3390 metres) rising above the other peaks away to the east in the boundary range between France and Italy, followed by a breathtaking, freezing, eight kilometres, multi-hairpin descent, losing more than six hundred and thirty metres of hard-won altitude on the way down to St Paul-sur-Ubaye, followed by a further seven kilometres descent through weird slate defiles to the N100, which is the main road from France to Italy by way of the Col de Larche. In the course of this descent from the Col de Vars I saw only one other human being: an elderly cyclist, like me *en tourisme*, pushing his laden bicycle fifteen kilometres uphill to the Col de Vars in carpet slippers. Very hot uphill stretch of eleven kilometres, at first through a dangerous-looking region in which all the rocks seem to be falling to pieces above the Ubayette River to Larche, where the French customs man didn't even bother to look up when I passed. Then a long, long climb through high pastures, the weather becoming colder and colder and more and more windy as I approached it, to the Col itself, a terrible sight with freezing cloud streaming over it, the wind so strong that on the way up to it I was blown off my bicycle and forced to walk the last kilometre or so – the first hill since leaving Wimbledon I hadn't ridden. Visibility nil at the frontier – which is 1994 metres up – and after attempting to attract the attention of a solitary, demoralized *carabinieri* sitting in a hut, whom I thought might offer me some sort of refreshment and, him failing to do so, I began the descent into Italy with chattering teeth, down an interminable, brutally pot-holed road in streaming rain to Pietraporzio on the River Stura, where I was given a warm welcome at the Albergo Regina delle Alpi (which was not much to look at from the outside but inside was filled with local people), *pasta e fagioli* (pasta made with beans) and lots of good wine from the Langhe in the plains of Piemonte, all for L1800 [90p or $2.15].

7 p.m. Reached Borgo San Dalmazzo after pedalling steeply downhill for fifty kilometres into the teeth of pouring rain and wind from the east so strong that I had the illusion of pedalling uphill. Put up at the Albergo Barra di Ferro a what was by then in Italy almost extinct sort of caravanserai with a galleried courtyard, having covered one hundred and forty kilometres, as many another wet and weary traveller arriving from France by way of the Col de Larche must have done. It also had a wonderful cook. Dinner L2000; room L1500.

Day 9

Still raining but less wet than Day 8. To Cuneo which had fearful, metal projections in the streets, manhole covers which had been given a shot of fertilizer? Death to cyclists whatever their function. Perhaps that was what was intended.

Italians are nice to cyclists in conversation; entombed in their vehicles they become brutes – especially the lorry drivers in their stinking, kilometre-long diesels who simply blast one into the ditch. After vile, twenty-five-kilometre journey to Fossano, unable to stand them any more, I abandoned major roads for more vegetable ones and wound my way up in low gear through the Barolo, the wine country in the foothills of the Langhe, to Alba, tough but quiet going through country in which white truffles are found. Very good, inexpensive and therefore truffleless lunch at the Gallo d'Oro in Albi – the best plan in Italy is to approach a well-fed-looking man in the street and ask him for *un ristorante dove si mangia bene e si spenda poco – alla casalinga* (a restaurant where one eats well and spends little – with home-cooking), a demand which rarely fails to elicit the name of at least a couple of good ones. Then through the Monferrato vineyards to Nizza Monferrato *ristorante alla casalinga*, altogether one hundred and twenty hard kilometres. Stayed the night at Da Italo in the market square. Ate the best *manzo bollito con salsa verde* (boiled beef with green sauce made with parsley, etc.) I have ever been offered and drank *dolcetto* and *barbera*, local red wines. Got rather drunk trying to decide which I liked best, but felt I deserved to do so after my exertions. Meal L2000, including two bottles of wine; room L1000.

Day 10

Without any trace of a hangover, which showed that Da Italo's wine had not been messed about with, raced thirty kilometres to Alessandria, which is half way between Turin and Genoa, in two hours, where I tried to have two broken spokes in the back wheel replaced, the lack of which was causing it to wobble in an alarming fashion. This involved removing the five-sprocket freewheel block but according to the mechanic – himself a racing cyclist – being a French block there was no local instrument that could accomplish this otherwise simple feat. However, being Italian, he replaced the spokes, which was very difficult with the block *in situ*, just the same.

Because of all this I missed the train and so rode on to Novo Ligure, thirty kilometres to the south, and took a train to Sarzana, the next stop after La Spezia before Carrara, this time without my bicycle which was sent off on another train and arrived half an hour after I did.

From Sarzana it was only about ten kilometres and twenty-six hairpin bends up to my destination near Fosdinovo, and when I negotiated the final bend to the place where the house was, I found that Wanda had arrived less than half a minute before, having left Wimbledon in a Land-Rover two days previously.

Altogether I had ridden about eleven hundred kilometres and I had lost three kilos in ten days. During this ride I had been terrified to remove the piece of sticking plaster I had stuck on my bottom at Sully-sur-Loire on Day 4, fearing that some large septic crater might have developed; but when I finally plucked up sufficient courage to peel it off, underneath there was nothing at all except undamaged skin. Did I imagine those happenings near Sully-sur-Loire?

34

Port-au-Prince
(1972)

'*Bon chance*,' said the chic air hostess. She had looked after us en route from Martinique where we had joined the plane. Now, she seemed genuinely reluctant that we should disembark at Port-au-Prince.

Thinking that we might have more *bon chance* without them, I had already got rid of the military maps of the island which I had bought before leaving London, although they were on such a small scale that they would only have been useful during a global war. With them, in the seat pocket, with the instructions of how to comport yourself in an emergency, the menu and the air sickness bags, I also left *The Comedians* by Graham Greene – whatever change the demise of Papa Doc might have brought about (and it was almost inconceivable that anything could be for the worse), I did not imagine that Greene could possibly yet be *persona grata* here. For the first time in my life I found myself reluctant to leave an airliner. At this moment to me it seemed a most desirable part of metropolitan France.

Outside on the tarmac, although half a gale was blowing, the heat was unbelievable. Borne on the fearful blast was the sound of welcoming music, produced by half a dozen Haitians, a rhythm with a strange resonance as if some of the musicians were armed with outsize jew's-harps. This was the *meringue*, a Creole rhythm and one that arriving and departing passengers were treated to in order to raise their spirits.

Inside, we joined a queue which eventually brought us to a desk at which was perched a large man of about fifty wearing dark glasses and an unseasonably thick Cheviot suit. It required no imagination to identify this senior citizen as a former member of Les Volontaires de la Sécurité Nationale, otherwise the Tonton Macoutes, the Bogeymen, an organization now, just over a year after Papa Doc's timely demise, still under a temporary cloud. (A homecoming, rather rash Haitian ahead of me in the queue had already expressed his feeling by spitting noisily and symbolically on the floor.)

'Visitor's card two US dollar! [78p]'

I told him that we had French francs, English pounds or British West Indian dollars but no US dollars.

'No! Two US dollar!'

Next to him, prominently displayed, was a notice board bearing the words: 'Business men! Make your stay in Haiti profitable by putting on a business!'

'Can you change a sterling traveller's cheque? American Express. Very good.'

'No! Two US dollar.'

Nor were *gourdes* any good, although *gourdes* were the local currency, with an even exchange rate with the dollar of twenty cents to one *gourde*. Not that I had any *gourdes*.

Eventually I persuaded this creature which, apart from its dark glasses, almost bullet-proof suit and limited gift of tongue, resembled one of the larger primates, to allow me through the barriers and through the chaos of the customs hall (where returning natives were being given a terrible going over by the officials), to cash a cheque.

It was a Sunday and there was no one to do it; but eventually an able fellow persuaded a taximan, who was hovering there, to give me some dollars in exchange for pounds, which he did at an appallingly adverse rate.

'You need a taxi,' said the taximan who had just done such a good deal, in American.

'I've ordered a car from Hertz,' I said rather smugly. I was already a bit fed up with Haiti.

'No Hertz car here today,' he said triumphantly.

I didn't believe him.

Back in Immigration I gave the Tonton Macoute four US dollars. He put the notes in an inside pocket, not in the suitcase in which I had watched him deposit most, but not all, of the money he received from the other temporary immigrants. Now, he showed signs of leaving.

'Hey!' I said, indignantly, 'what about my visitor's card?'

He gave me the sort of inscrutable look which you only really get from people who wear dark glasses and vanished through a door marked 'No Admittance'.

By the time we reached the customs hall the customs officers had made off too. No need to have jettisoned my maps or the work of the master, Greene.

Just as the taximan said, the car I had ordered by cable from Martinique was not there.

Knowing, because they had been so boringly reiterative about it, that they have to try harder because they are only 'Number Two', I got through to Avis and a voice promised me a Volkswagen in half an hour. Triumphantly, I relayed this information to the taximan. It was now a quarter past twelve.

'All right,' he said, 'just you see. It won't come till three o'clock and Sunday it may never come. Just you see.'

By a quarter to two we were beginning to feel that both the taximan and Al Seitz, the proprietor of the Grand Hotel Oloffson, were probably right. In spite of his name Seitz had sounded like a quiet American on the telephone.

'Don't wait for Hertz or Avis or anything else,' he said. 'Get a taxi. Right now we're taking the children to the beach for the afternoon but when you get here tell César, the barman, to pay the taximan if you haven't got any cash. If you do pay him only give him $3.50 [£1.50] whatever he says.'

Outside, in the featureless plain, in which the Aeroport François Duvalier stands between the mountains and the invisible sea, the atmosphere was incandescent, with the brightness of a well-tended gas mantle. Beyond, the mountains looked as if they were about to explode. Occasionally, running downwind among scurries of dust, emaciated donkeys trotted past, each with a number of persons on board, on their way to some belated assignation in Port-au-Prince.

We passed the time reading a coffee-table book which I found abandoned in the deserted customs hall: *Haiti. The First Negro Republic in the World. Its True Face*. Printed in Belgium, as the blurb said, 'by Private Initiative'.

The one who had provided the private initiative had now gone, but what would be, except for the association of ideas, his totally unmemorable visage still peered out in a colour photograph from page thirteen, with a framed portrait of Pope

Paul VI on the mantelshelf behind him, and a selection of UN flags. Papa Doc, Eighth President for Life, now officially known as 'Le Grand Disparu'.

'In the traits of dignity and pride of the Head of the Nation, the concentration of the thinker, the impassibility of the conqueror, the true vision of the Haitian country,' ran the Mr Toad-like caption. Looking at the photograph, it was difficult to imagine that he had had the head of an opponent hewn from his shoulders, flown to him from an outlying *département* in a bucket of ice, placed in a deep freeze in his palace and, from time to time, had it brought to him after office hours so that he could contemplate it in his in-tray during the watches of the night.

And there was a photograph of the Eighth President for Life's praetorian guard, the Tonton Macoutes, some eight thousand of whom were said now to be building roads in remote places. Not all, however, 'Two US dollar' could scarcely be said to be in disgrace, nor Luckner Cambronne, the power behind the new president and one of the most corrupt men in this most corrupt of islands who had raised the rake-off to the level of pure art. Here, in this out-of-date book, a detachment of them was shown paraded in their bright blue uniforms, rarely worn for the dreadful tasks which they performed with such assiduity because such things were more easily carried out in the anonymity of the dark suits which were their everyday working uniforms (together with the dark glasses), on which bloodstains were less evident than on bright blue.

After a discussion about the fare, difficult to maintain in such heat, on an empty stomach and because his was the only remaining taxi, we entered 'my' taximan's taxi, a huge, black, hearse-like Steinbergian vehicle, and were driven off down an avenue which seemed to have been intended as a triumphal way but lacked any of the panoply of triumph, through some unmemorable outskirts and into the Avenue Dessalines, otherwise La Grande Rue, which runs through the heart of the city, leaving on the right hand La Route du Fort Dimanche, the fort itself marked on an oil company's map (free to visitors) as '*Lieu d'Intérêt*'. Interesting, presumably, because few prisoners had ever emerged from it in one piece and because it was to this place, in the middle of the night, that le Grand Disparu took his son-in-law, the husband of his eldest daughter, Marie-Denise (herself later a contender for supreme power), to witness, as a mark of his disapproval of her marriage to a potentially dangerous army colonel, the execution of nineteen of his friends by an impromptu firing squad composed of staff officers. At the last moment, having accepted a mandatory invitation to be present, the officers had rifles thrust into their hands and were ordered to shoot their companions in arms, who were lashed to stakes.

The taxi droned on into the city.

We sat on the veranda of the Grand Hotel Oloffson, having consumed a delicious light tropical luncheon, served by an old retainer who answered to the name of C'est Dieu. As we found it impossible either to hail him by this name – or even to paraphrase it, one could scarcely call a waiter God, even in Haiti – we contented ourselves with attracting his attention in the English manner, that is by raising a hand weakly and calling out, 'Er . . .' or 'I say . . .'

This was the hotel of *The Comedians* and from where we were sitting, down in a corner of the tropical garden where huge, rather rickety palm trees soared into the sky, we could see the swimming pool in which, empty in the novel, some minister or other had messily done away with himself. Now, full of water, it looked positively inviting. Meanwhile, overhead in the trees, strange birds uttered rasping cries.

The Grand Hotel Oloffson stood on the lush, lower slopes of Kenscoff Mountain which, particularly towards evening, loomed over the city in a somewhat alarming manner, as if it was about to fall and squash it flat. The hotel was a truly astonishing structure in a country in which this epithet has become meaningless from sheer over-use. Painted a sizzling white, it had been built of almost indestructible mahogany to house Simon Sam, who was President from 1896 to 1902, not to be confused with his namesake, Guillaume Sam, also President (though a short-term one), who ended up being impaled by a mob on the railings of the French Embassy in 1915, after which he was torn to pieces.

It was the embellishments that made it unique. From every possible and impossible vantage point it sprouted turrets, spires, crotchets, finials and balconies, some of which appeared to have been put on upside down, all of them riddled with so much fretwork that it was a miracle that the building remained standing. It was as if some giant, but inspired, wood-boring insect had been let loose on what had originally been a decent, solid, colonial, clapboard structure. What it was, although it lacked a dome – about the only thing it did lack, except flying buttresses – was oriental, the sort of orient suggested in Coleridge's opium-induced visions of Xanadu. The bar with the brilliant colours of the primitive Haitian paintings shining through the perpetual dusk of daytime, the dining-room with its rocks and greenery could well have been the ante-chambers to caverns leading down to a sunless sea, which was exactly what guests craved for after a morning's sightseeing in the inferno of Port-au-Prince.

'The darling of the literary and theatre sets,' someone said or wrote of the Oloffson – and it was true. Even a short list of personalities who had stayed there read like the formula for some deadly gas; and in what I would have thought an unguarded moment Al Seitz had had this thought for the day inscribed on the postcards which I could send home, with a picture of the hotel on the front to Three Ther Mansions, which my mother though in failing health still kept on after my father's death. Perhaps it was not so silly, but whether it was or not did not matter because, in Haiti, communications from foreigners rarely, if ever, left its shores – not because of censorship but because the stamps necessary to export a letter were of such high denomination that as soon as they entered the sorting offices the impoverished staff immediately steamed them off.

Above my head on the walls of the balcony were a number of primitive paintings, robust enough to be exposed to the almost open air, some of them excellent. I was particularly attracted by one painted by an artist whose name was A. Bazile. In marvellous colours – gold and cobalt and splendid reds – it showed a number of giant peasants, the women with great baskets of Indian corn on their noddles standing in front of what was, by a trick of the perspective, a diminutive village.

There were no guests about and all of the twenty-eight servants, which Seitz's brochure assured me existed, had retired for the siesta, including C'est Dieu. Wanda had gone to unpack.

'Est-ce-que vous aimez ce tableau?'

Far below me – I was standing on a chair pondering the problem of taking a colour photograph of this painting *contre-jour* – was a solemn-looking, owl-like, very, very, black Negro.

'Oui, c'est très, très bon.'

'Twès bien! Excellent. Je suis Bazile.'

He spoke with the omission of the 'r' sound, something which all Créoles affect. It sounds attractive on a woman's lips, a little soppy, a little Woosterish, on a man's.

I was delighted to meet him – my first primitive painter, and all to myself. Exclusive! We ended up by having a long conversation on the balcony of our room while I fed the three of us rum punches. He told us the kind of paintings he liked to paint and the materials he employed. His father, he said, had been the great Bazile whose murals form part of the famous decoration of the Episcopalian Cathedral of the Holy Trinity in Port-au-Prince. In one mural, *The Last Supper*, by another Haitian painter named Bottex, Judas is a white man in an otherwise all-black cast.

'I am not as good as my father,' he said, modestly.

We talked of the foreigners he had met.

'Vous connaissez, naturellement, Connolee?' he said.

I said that I did not naturally know Cyril Connolly, but that I had read his books, had bought his magazine *Horizon* in 1939 and 1940, and that he contributed to the *Sunday Times* and was famous.

He told me that he had always wanted to have one of his paintings reproduced in a magazine in colour.

'If I painted one would you have it produced in your magazine [*The Observer Magazine*]?' he asked. 'Of course it would be completely free.'

I insisted on paying, but he would not hear of it.

'It will be slow coming because I am a slow painter,' he said, 'and because it is too expensive to send these pictures with their frames by air.'

I asked him how much it would cost to send it by surface mail. He said about $12 (£4.60). By this time I had acquired a $20 bill but I still had no smaller change and because I did not want to embarrass M. Bazile, now negligently sketching the sort of daisy in my notebook that I would have drawn if I had been called upon to sketch a daisy, and who knew M. Connolee, I gave him $20 (£7.60).

When I told Al Seitz back from the beach with his wife and children he roared with laughter. I had imagined him from his voice on the telephone as being about thirty and outfitted by Brooks Brothers. He turned out to be a cigar smoking presence of fifty-plus. 'That must be the Tonton Macoute who used to keep tabs on the hotel. Solid bone from the feet up. Couldn't spell his own name let alone write it. About the only thing he could paint would be a hearse. He's coming on. It's obviously doing these boys a lot of good having to fend for themselves.'

Before we left Port-au-Prince we bought two of the real M. Bazile's paintings. We also visited, out of curiosity, the *rue* in which the *faux* Bazile said he lived – a pitch-black lane full of potholes in a weird suburb to the west of the city beyond the red-light quarter – only he and his house were missing.

Behind the hotel was a ramshackle, circular construction with a corrugated-iron roof which looked like an open umbrella, below which we could see hundreds of black and off-black feet dangling. This was a *gaguère*, a cockpit, and the owners of the feet were the spectators. Inside, it was like an engraving by Hogarth, but one in which all the protagonists were black. Huge sums of money were changing hands; *clairin*, the cheapest and rawest spirit, was circulating; the cocks were being anointed with rum and having their feathers dampened by their owners, who filled their mouths with water and squirted it on to the birds, which were also being admired and criticized by the punters; and the air was filled with the sound of their crowing. There were no women present, and although I was the only white man, the atmosphere was distinctly friendly.

Soon a main began on a floor of beaten earth. The birds did not wear steel spurs. There was no need; their own were sharpened so that they were like stilettos. I knew from reading accounts of the *gaguère* by other visitors to Haiti that I was not going to enjoy it, and when the main reached a point when one bird had had one of its wings partly torn off and the other was eyeless I left.

295

That evening, and every evening, César the barman began circulating the rum punches around six-thirty: to those members of the literary and theatre sets who called the place darling; to exquisite and beautiful members of the local smart set whose patrician countenances ranged in pigmentation from the jet-black of Africa to the palest of pale colour of the mulatto *sangmêlées*, down from the heights of Pétionville for a night out (nowhere else in the Caribbean, rich or poor, could you see such beautiful people). And there was M. Aubelin Joliecoeur – the Petit Pierre of Greene's novel – a tiny, posturing, chocolate-coloured journalist and PR man to the regime, who must have been more clever than he looked or sounded to have survived at all, although according to those who disliked him, of whom there were large numbers, he was by then down to his last white silk suit. There he was calling people he had never seen before 'darling', kissing their hands, waving a little cane, popping in and popping out again to bully his wretched, abject chauffeur, here, there and everywhere, night and day, all over the town; and there were two leathery US Marines in sweaty civilian suits, over from Florida to train *les Léopards*, the new regime's successors to the Tonton Macoutes. In addition, there were a number of more conventional visitors, of whom we were a pair, who looked at the other guests, the gleaming paintings, the whole exotic set-up, with frank curiosity and ate the delicious, outlandish food with relish. And presiding over this motley lot was Al (Garcia Vega) Seitz, who had bought the place from a Frenchman without descending from the taxi to view it because it was raining at the time, and his willowy wife Sue, from Bucks County, Pennsylvania, whom he had married in the bar using a cigar band for a ring because he had mislaid the real one.

These Firbankian impressions (did Firbank really say 'They tell me the president's a perfect dear' before setting out for Haiti?) persisted into the early hours of the following morning, which were the most supportable of the day in Port-au-Prince, when, through a frame of tropical vegetation, we saw the cupolas of the cathedral rising from the mist that covered the city; listened to the bells calling the people to early mass; heard the jaunty trumpet calls from the Caserne Dessalines in which so many unsuccessful coups had been either hatched or nipped in the bud, and the distant roar of *camionettes*, the gaudily-painted vehicles of the public bus system with names such as Sauve-qui-peut, Toujours Immaculée and Grace-à-Dieu, as they thundered through the town, at ten cents a ride – apart from shared taxis, the *camionettes* were the only way of getting about the place without being skinned and by far the best way of meeting the inhabitants.

Later, as it grew progressively hotter, we walked down into Port-au-Prince through lanes flaming with bougainvillea in which passers-by said in a friendly way, '*Bonjour, blancs*' – a pleasant change from some other, at that time, British islands in the Caribbean, such as St Lucia, where we had been told: 'Go home, whites.'

On the way into Port-au-Prince we saw some of the principal sights: the tomb of le Grand Disparu, what looked like a *bijou* villa in an enormous, dazzlingly white cemetery, guarded by sentries who also looked after the cylinder which provided gas for the perpetual flame; and an enormous hoarding on top of a grandstand on the Champ de Mars on which was displayed a several times larger than life figure of Jean-Claude Duvalier, the Doctor's obese son, Ninth President for Life, in full evening dress with the banner headline, 'L'Idole du Peuple'.

'*Il est un bon garçon*,' a loyal passer-by said as I stopped to photograph this extraordinary advertisement, '*un twès, twès, bon garçon.*'

On past the strategically situated Caserne Dessalines, next door to the palace, painted a rather overpowering shade of orange, with the sounds of a military band floating out through sun-blinded windows; on past the palace – no stopping here, not even for tourists (if you did a sentry shouted to you to keep moving) – none of the locals ever came here unless they were given some sort of inducement anyway; on into the dilapidated heart of the city by the Grand Rue now, at this hour, with the iron riot doors of its shops flung open, the arcades in front of them jammed with street tradesmen, the air in this heat lethally heavy with the exhaust of a thousand deep-loaded *camionettes*; on as far as the Iron Market built in 1889 by President Hyppolite, one of Haiti's more distinguished presidents, by this time having been asked by every second person in the Grand Rue, young and old, to give them one dollar, but with a complete lack of hope of receiving anything at all, however little. These were not professional beggars – the professional beggars, including the mutilated ones, had been whisked away, one wondered where. They had been whisked away before at the whim of successive ministers of tourism. They always came back, or others like them. Those who had temporarily taken their place were just ordinary people, desperately poor.

On sale in the market were pretty but difficult to transport lampshades made of shiny tin, clay pipes which were smoked by women in the rural interior, locally-made brassieres – brassieres, baseballs, textiles, 'quickie' divorces and blood plasma bought from the already anaemic inhabitants at $3 (£1.15) a litre were some of the most thriving exports to the United States from a country in which the legal wage of a manual worker was $1 a day (no wonder American big business was attracted to Haiti) – and there were oleographs of the Virgin of the Seven Sorrows with a bare and transfixed, blood-red heart, which in voodoo is the emblem of Erzulie Freda, the Dahomeyan Goddess of Love and of St Rose of Lima, St James the Major and St Charles Borromeo, just a few of the Christians who are also part of the voodoo pantheon and hang above their altars in the *tonnelles*, the voodoo peristyles.

Up the hill, by the cathedral, poor women, dressed in white, indistinguishable from the ecstatic devotees who crowd the voodoo *tonnelles* at night, knelt on the pavement before the shrines with their arms outstretched imploringly or else they clung to the railings as if the Devil was trying to drag them away. Or was it Baron Samedi, Lord of the Cemeteries and Chief of the Legion of the Dead, in his frock coat, bowler hat and carrying a black walking stick? To whom were those women addressing themselves – the Christian Trinity and the Saints, or the Gods of Africa; Bon-Dieu-Bon, otherwise Le Grand Maître who is sometimes male and sometimes female; Ogoun Feraile, God of War; Bossou Comblamin; or the Virgin of the Seven Sorrows or St Rose of Lima, in their other guises? In Haiti Catholicism and voodooism are so inextricably mixed in the minds of the people, while the Houngans, the voodoo priests, take what they want from the Church that many Catholic priests, after strenuous but totally unsuccessful attempts to destroy voodooism by cutting down sacred trees and destroying the complex apparatus of worship, have given up in despair trying to disentangle one from the other.

From the cathedral it was only a short distance to a shanty town in which the houses were little more than packing-cases, the corrugated-iron roofs by day almost red hot, the streets narrow ditches. Yet here, where the poorest of the urban poor lived and where the *tonnelles* were thickest, visiting them at night you were, according to the local inhabitants, safe.

It was a pity that the time we spent in Haiti coincided with Easter, a close season for genuine voodoo, and the drums were silent throughout the island. Instead, what were known as Raras, wild-looking bands of revellers carrying kerosene lanterns and playing long bamboo pipes roamed the streets and lanes, each with a Rara king dressed in a tunic embroidered with thousands of sequins, the Haitian equivalent of Pearlies.

Although it was the close season for voodoo we visited one of the *tonnelles* on the outskirts of the city. Our guide was the Houngan, a tall, thin, praeternaturally intelligent-looking man who spoke only Créole. The peristyle was open-sided and had a palm-leaf roof supported at the centre by a painted and decorated pole which played an important part in the ceremonies. In huts round about there were altars and all the complex apparatus of voodoo: bottles filled with strange substances, swords, crucifixes, bells, anthropomorphic paintings of gods and goddesses, gourds enclosed in beads and vertebrae, china pots that looked as if they ought to have contained *foie gras*, also tied up with beads, pincers, iron serpents, old bedsteads and, on the walls, mystical patterns and pictures of the Virgin and the Saints who had a place in the ceremonies. In the smallest hut, Le Caye Zombi, there were shackles and whips. And there was, of course, the complete equipment of The Baron, set out like a gentleman's wardrobe, ready for the day, but on a black cross. It was all very interesting, but it needed the ecstatic participants in the rites to give it any meaning.

35

Leaving *The Observer*
(1973)

In the autumn of 1973 I left *The Observer*, but only after much heart-searching and against the advice of Donald Trelford, who subsequently became editor, and many other friends who worked for it. It was the only job I ever had in the whole of my life that I was genuinely sorry to leave and I still continue to write for the paper from time to time. I gave it up because in the nine years I had been its travel editor the mechanism of travel had changed out of all recognition.

The great majority of travellers, myself included, were now moved around the world en masse, rather like air freight, and just like freight when they reached their destinations they were lifted out of the bowels of the aircraft and delivered to their hotel rooms. It was not that people really wanted to be treated like this but this, it was emphasized, was the most economical way of travelling, and it was largely true.

'I suppose there ought to be a staircase,' Evelyn Waugh makes Professor Silenus say gloomily in *Decline and Fall*, contemplating the immense country house he is constructing for Mrs Beste-Chetwynd. 'Why can't the creatures stay in one place? Do dynamos require staircases? Do monkeys require houses?'

By the 1960s no such residual doubts clouded the minds of those who designed and furnished hotels. By then a staircase was something put in to satisfy some regulation. Movement from one floor to another was effected in hot little, neon-lit boxes which were filled with Muzak unless the current failed in which case the occupants were marooned indefinitely.

The new hotels were built of concrete. Expensive ones were unnecessarily solid like a blockhouse in the Atlantic Wall built by Todt slaves. If less expensive they shook like a jelly when one got into bed and you could hear the most unbelievable things going on all around you. There was the room itself, with its view or non-view through what are intended to be permanently sealed windows, its walls adorned with absolutely characterless pictures. And there was the lighting, which was impossible to read by. Indeed, the lighting's only function seemed to be to cast, if you had someone to cast it with, an erotic shadow. It was certainly impossible to write at the attenuated, so-called desk. Round the corner, in the even more attentuated bathroom, was a bath so short that somewhere, one felt, there must be a circular saw with which to convert oneself into two or more submersible parts. Next to it was the lavatory basin, its seat decorated with what looked like a drum majorette's sash, bearing the legend 'It's sanitized', which suggested the possibility that nameless acts may have been performed in revenge by those who had to drape it in this demeaning fashion. And downstairs was the restaurant with its 'international specialities' and pastiches of local 'specialities', in which no local ever ate unless trying to conclude some deal with a visitor.

Entombed in such places I thought with nostalgia of Japanese inns in which some of the pleasures were decidedly unexpected. The silence of the Pera Palace, shattered only once it is said by a Bulgarian whose suitcase, full of bombs, blew up in the hall as he was registering; the romantic decrepitude of the Bela Vista at Macau; the sleaziness of the Cavendish in the days of Rosa Lewis; the inspired improbability of the Oloffson at Port-au-Prince; and the friendliness of a certain pub on the estuary of the Kenmare River. I felt, too, and I felt myself responsible having for years written about lonely places, that the time was not far off when there would be no place on earth accessible to ordinary human beings in which they would be able to feel themselves alone under the sky without hearing the noise of machines.

By far the greatest menace to the lonely places was the bulldozer. With the bulldozer roads could be made, through the wilderness and over mountain ranges, in a few months, which would previously have taken years to construct and would never have been built at all because of the cost. Most of these so-called 'panoramic roads' were not intended for the convenience of the inhabitants. They were made for tourists in motor cars who never got out of their vehicles at all. No one who lived in a remote place and enjoyed doing so was safe from the panoramic road. By 1973 they had already destroyed the solitude of the high Apennines which I knew and loved so well.

Even worse will be the day, which has not yet come, when the desire to be alone has finally been extinguished from the human heart.

Acknowledgements

The letters from Evelyn Waugh on pp. 170, 171 and 172 are reproduced by permission of A. D. Peters and Co. Ltd.

I would like to take this opportunity to thank Robin Baird-Smith and Gill Gibbins of Collins for their advice and help. Also Ann Etherington for her accurate and unbelievably rapid typing of the manuscript not once but several times before it was reduced to publishable proportions.

P083182 Inv. 1827847